12th International Workshop on Health Text Mining and Information Analysis (LOUHI 2021)

Online
19 April 2021

ISBN: 978-1-7138-2969-0

Printed from e-media with permission by:

Curran Associates, Inc.
57 Morehouse Lane
Red Hook, NY 12571

Association for Computational Linguistics
209 N. Eighth Street
Stroudsburg, Pennsylvania 18360

Phone: 1-570-476-8006
Fax: 1-570-476-0860

acl@aclweb.org

Additional copies of this publication are available from:

Curran Associates, Inc.
57 Morehouse Lane
Red Hook, NY 12571 USA
Phone: 845-758-0400
Fax: 845-758-2633
Email: curran@proceedings.com
Web: www.proceedings.com

EACL 2021

The 12th International Workshop on Health Text Mining and Information Analysis LOUHI 2021

Proceedings of the Workshop

April 19, 2021

Introduction

The International Workshop on Health Text Mining and Information Analysis (LOUHI) provides an interdisciplinary forum for researchers interested in automated processing of health documents. Health documents encompass electronic health records, clinical guidelines, spontaneous reports for pharmacovigilance, biomedical literature, health forums/blogs or any other type of health-related documents. The LOUHI workshop series fosters interactions between the Computational Linguistics, Medical Informatics and Artificial Intelligence communities. The 11 previous editions of the workshop were co-located with SMBM 2008 in Turku, Finland, with NAACL 2010 in Los Angeles, California, with Artificial Intelligence in Medicine (AIME 2011) in Bled, Slovenia, during NICTA Techfest 2013 in Sydney, Australia, co-located with EACL 2014 in Gothenburg, Sweden, with EMNLP 2015 in Lisbon, Portugal, with EMNLP 2016 in Austin, Texas; in 2017 was held in Sydney, Australia; in 2018 was co-located with EMNLP 2018 in Brussels, Belgium; in 2019 was co-located with EMNLP 2019 in Hong Kong; and in 2020 was co-located with EMNLP 2020 and took place online due to the COVID-19 pandemics. This year the workshop is co-located with EACL 2021 and takes place online due to the persistence of the COVID-19 pandemics.

The aim of the LOUHI 2021 workshop is to bring together research work on topics related to health documents, particularly emphasizing multidisciplinary aspects of health documentation and the interplay between nursing and medical sciences, information systems, computational linguistics and computer science. The topics include, but are not limited to, the following Natural Language Processing techniques and related areas:

- Techniques supporting information extraction, e.g. named entity recognition, negation and uncertainty detection
- Classification and text mining applications (e.g. diagnostic classifications such as ICD-10 and nursing intensity scores) and problems (e.g. handling of unbalanced data sets)
- Text representation, including dealing with data sparsity and dimensionality issues
- Domain adaptation, e.g. adaptation of standard NLP tools (incl. tokenizers, PoS-taggers, etc) to the medical domain
- Information fusion, i.e. integrating data from various sources, e.g. structured and narrative documentation
- Unsupervised methods, including distributional semantics
- Evaluation, gold/reference standard construction and annotation
- Syntactic, semantic and pragmatic analysis of health documents
- Anonymization/de-identification of health records and ethics
- Supporting the development of medical terminologies and ontologies
- Individualization of content, consumer health vocabularies, summarization and simplification of text
- NLP for supporting documentation and decision making practices
- Predictive modeling of adverse events, e.g. adverse drug events and hospital acquired infections
- Terminology and information model standards (SNOMED CT, FHIR) for health text mining

- Bridging gaps between formal ontology and biomedical NLP

The call for papers encouraged authors to submit papers describing substantial and completed work but also focus on a contribution, a negative result, a software package or work in progress. We also encouraged to report work on low-resourced languages, addressing the challenges of data sparsity and language characteristic diversity.

This year we received 20 submissions. Each submission went through a double-blind review process which involved three program committee members. Based on comments and rankings supplied by the reviewers, we accepted 11 papers. The selection was entirely based on the scores provided by the reviewers. The overall acceptance rate is 55%.

Our special thanks go to Karin Verspoor for accepting to give an invited talk.

Finally, we would like to thank the members of the program committee for providing balanced reviews in a very short period of time, and the authors for their submissions and the quality of their work.

Organizers:

Alberto Lavelli (FBK, Trento, Italy)
Eben Holderness (Brandeis University, USA)
Antonio Jimeno Yepes (IBM Research Australia)
Anne-Lyse Minard (LLL, CNRS, University of Orléans, France)
James Pustejovsky (Brandeis University, USA)
Fabio Rinaldi (Dalle Molle Institute for Artificial Intelligence Research - IDSIA, Switzerland & FBK, Trento, Italy)

Program Committee:

Mohammad Akbari, National University of Singapore, Singapore
Rafael Berlanga Llavori, Universitat Jaume I, Spain
Georgeta Bordea, Université de Bordeaux, France
Leonardo Campillos Llanos, LIMSI, CNRS, France
Francisco Couto, University of Lisbon, Portugal
Hercules Dalianis, Stockholm University, Sweden
Kerstin Denecke, Bern University of Applied Sciences, Switzerland
Natalia Grabar, CNRS UMR 8163, STL Université de Lille3, France
Cyril Grouin, LIMSI, CNRS, Université Paris-Saclay, Orsay, France
Thierry Hamon, LIMSI, CNRS, Université Paris-Saclay, Orsay, France & Université Paris 13, Villetaneuse, France
Aron Henriksson, Stockholm University, Sweden
Eben Holderness, Brandeis University, USA
Rezarta Islamaj-Dogan, NIH/NLM/NCBI, USA
Antonio Jimeno Yepes, IBM Research, Australia
Yoshinobu Kano, Shizuoka University, Japan
Jin-Dong Kim, Research Organization of Information and Systems, Japan
Dimitrios Kokkinakis, University of Gothenburg, Sweden
Martin Krallinger, Spanish National Cancer Research Centre (CNIO), Spain
Alberto Lavelli, FBK, Trento, Italy
Analia Lourenco, Universidade de Vigo, Spain
David Martinez, University of Melbourne and MedWhat.com, Australia
Sérgio Matos, University of Aveiro, Portugal
Timothy Miller, Harvard Medical School, USA
Anne-Lyse Minard (LLL, CNRS, University of Orléans, France)
Hans Moen, University of Turku, Finland
Diego Molla, Maquaire University, Australia
Danielle L Mowery, University of Utah, USA
Aakanksha Naik, CMU, USA
Aurélie Névéol, LIMSI, CNRS, Université Paris-Saclay, Orsay, France
Mariana Lara Neves, German Federal Institute for Risk Assessment, Germany
Jong C. Park, KAIST Computer Science, Korea
Laura Plaza, Universidad Complutense de Madrid, Spain
James Pustejovsky, Brandeis University, USA
Fabio Rinaldi, Dalle Molle Institute for Artificial Intelligence Research - IDSIA, Switzerland & FBK, Trento, Italy
Thomas Brox Røst, Norwegian University of Science and Technology, Norway
Tapio Salakoski, University of Turku, Finland

Stefan Schulz, Graz General Hospital and University Clinics, Austria
Maria Skeppstedt, Linneus University, Sweden, and Potsdam University, Germany
Amber Stubbs, Simmons College, USA
Hanna Suominen, Australian National University, Australia
Suzanne Tamang, Stanford University School of Medicine, USA
Özlem Uzuner, MIT, USA
Yanshan Wang, Mayo Clinic, USA
Pierre Zweigenbaum, LIMSI, CNRS, Université Paris-Saclay, Orsay, France

Additional Reviewers:

Fatemah Husain, Kuwait University (Kuwait)

Invited Speaker:

Karin Verspoor, RMIT University (Australia)

Table of Contents

ArCorona: Analyzing Arabic Tweets in the Early Days of Coronavirus (COVID-19) Pandemic

Hamdy Mubarak, Sabit Hassan
Qatar Computing Research Institute
HBKU, Qatar
{hmubarak, sahassan2}@hbku.edu.qa

Abstract

Over the past few months, there were huge numbers of circulating tweets and discussions about Coronavirus (COVID-19) in the Arab region. It is important for policy makers and many people to identify types of shared tweets to better understand public behavior, topics of interest, requests from governments, sources of tweets, etc. It is also crucial to prevent spreading of rumors and misinformation about the virus or bad cures. To this end, we present the largest manually annotated dataset of Arabic tweets related to COVID-19. We describe annotation guidelines, analyze our dataset and build effective machine learning and transformer based models for classification.

1 Introduction

As the Coronavirus (COVID-19) crippled lives across the world, people turned to social media to share their thoughts, news about vaccines or cures, personal stories, etc. With Twitter being one of the popular social media platforms in the Arab region, tweets became a major medium of discussion about COVID-19. These tweets can be indicators of psychological and physical well being, public reactions to specific actions taken by the government and also public expectation from governments. Therefore, identifying types of tweets and understanding their content can aid decision making by governments. It is also important for governments to identify and prevent rumours and bad cures since they can bring harm to society.

While there have been many recent works about tweets related to COVID-19, there are a very few targeted toward aiding governments in their decision making in the Arab region despite Arabic being one of the dominant languages on Twitter (Alshaabi et al., 2020). Some of the existing works use automatically collected datasets (Alqurashi et al., 2020). Manually labeled datasets are either small in size (few hundred tweets) (Alam et al., 2020) or target a different task such as sentiment analysis (Haouari et al., 2020). To fill this gap, we present and publicly share the largest (to our best knowledge) manually annotated dataset of Arabic tweets collected from early days of COVID-19, labeled for 13 classes. We present our data collection and annotation scheme followed by data analysis, identifying trends, topics and distribution across countries. Lastly, we employ machine learning and transformer models for classification.

2 Related Work

Much of recent works on COVID-19 rely on queries to Twitter or distant supervision. This allows a large number of tweets to be collected. Chen et al. (2020) collect 123M tweets by following certain queries and accounts on Twitter. GeoCoV19 (Qazi et al., 2020) is a large-scale dataset containing 524M tweets with their location information. Banda et al. (2020) collected 152M tweets at the time of their writing. Li et al. (2020) identifies situational information about COVID-19 and its propagation on Weibo. Other works include propagation of misinformation (Huang and Carley, 2020; Shahi et al., 2020), cultural, social and political impact of misinformation (Leng et al., 2020) and rumor amplification (Cinelli et al., 2020).

For Arabic, we see a similar trend where few datasets are manually labeled. Alqurashi et al. (2020) provide a large dataset of Arabic tweets containing keywords related to COVID-19. Similarly, ArCOV-19 (Haouari et al., 2020) is a dataset of 750K tweets obtained by querying Twitter. Alam et al. (2020) annotate a small number of English (currently 504) and Arabic tweets (currently 218) for (i) existence of claim and worthiness of fact-checking (ii) harmfulness to society, and (iii) relevance to governments or policy makers. Yang et al.

Proceedings of the 12th International Workshop on Health Text Mining and Information Analysis, pages 1–6
April 19, 2021.

(2020) annotate 10K Arabic and English tweets for the task of fine-grained sentiment analysis.

Alsudias and Rayson (2020) collected 1M unique Arabic tweets related to COVID-19 between December 2019 and April 2020. They used K-means algorithm from Scikit-learn Python package to cluster tweets into 5 clusters, namely: statistics, prayers, disease locations, advising, and advertising. They also annotated random 2000 tweets for rumor detection based on the tweets posted by the Ministry of Health in Saudi Arabia.

3 Data Collection

We used twarc search API[1] to collect tweets having the Arabic word كورونا (Corona) in Feb and March 2020. We collected 30M tweets in total. The reason behind selecting this word is that it's widely used by normal people, news media[2] and official organizations [3] as opposed to كوفيد-19 (COVID-19) which is rarely used by normal people in different Arab countries based on our observations. We aimed to increase diversity of tweet sources. Our collection covers the period from Feb 21 until March 31 in which Coronavirus was reported for the first time in All Arab countries except United Arab Emirates (AE)[4] (Jan 29) and Egypt (EG) (Feb 14). The date of the first reported Coronavirus case in Lebanon (LB) was Feb 21, and in Iraq (IQ), Bahrain (BH), Oman (OM), and Kuwait (KW) was Feb 24, in Qatar was Feb 29, and in Saudi Arabia (SA) was March 2. All other Arab countries came later.

4 Data Annotation

During the period of our study (40 days), we extracted the top retweeted 200 tweets in each day (total of 8000). We assume that the top retweeted tweets are the most important ones which get highest attention from Twitter users. Annotation was done manually by a native speaker who is familiar with Arabic dialects according to class descriptions shown in Table 1. To measure quality, we annotated 200 random tweets by a second annotator. Inter-annotator agreement was 0.85 using Cohen's kappa coefficient which indicates high quality given that annotation is not trivial and some classes are close to each other. Examples of annotation classes are shown in Figures 1 and 2.

Note: If a tweet has multimedia (image or video) or an external link (URL or another tweet), the annotator was asked to open it and judge accordingly to consider the full context of the tweet. For example, if a tweet has a text about a prayer and the attached image is about number of new cases, this should be classified as REP not PRAYER.
Data can be downloaded from this link[5]: `https://alt.qcri.org/resources/ArCorona.tsv`.

4.1 Limitation

We found that ≈10% of the tweets can take more than one class, e.g. a tweet reports new cases and a medical advice. We plan to allow multiple labels in future. In the current version, such tweets take the label of the first "important" class. We consider the first 8 classes in Table 1 to be important and the last 5 classes (PRSNL, SUPPORT, PRAYER, UNIMP and NOT_ARB) to be less important. These classes will be merged into LessImportant class.

5 Analysis

Class timeline is shown in Figure 3. We can observe the following important notes:

- Large portion of tweets can be considered as LessImportant to many people (≈ 30%).
- Reports (REP) and actions taken by governments (ACT) are the most retweeted tweets.
- Information about the virus (INFO) get less attention with time and there is an increasing number of tweets about volunteering (VOLUNT).
- There are continuous requests for governments to take actions (SEEK_ACT) – especially in the beginning (≈ 15%), and few tweets are about rumors (≈ 5%) and cures (≈ 2%).

We took a random sample of 1000 tweets and annotated them for their topics. Figure 4 shows that, in addition to health, the virus affected many aspects of people's lives such as politics, economy, education, etc. We found also that 7% of tweets have hate speech, e.g. attacking China and Iran for spreading the virus as shown in Figure 5.

[1] https://github.com/DocNow/twarc

[2] https://www.aljazeera.com/topics/events/coronavirus-outbreak.html

[3] https://www.who.int/ar/emergencies/diseases/novel-coronavirus-2019

[4] We use ISO 3166-1 alpha-2 for country codes

[5] We share tweet id, date and class.

Class	Description	Count
1. REP	Reports and announcements such as number of infections, recovery cases and deaths.	1664
2. ACT	Measures or actions taken by governments such as curfew, closing of country borders, shops and worship places. This includes discussions and consequences of these measures.	1383
3. INFO	Information about the virus, symptoms, incubation period, how it spreads, mask types, etc.	300
4. RUMOR	Rumor or refute rumor. A rumor is a circulating story or report of uncertain or doubtful truth.	421
5. ADVICE	Advice such as washing hands, staying at home, wearing masks and avoiding travel.	1047
6. SEEK_ACT	Seek actions from governments such as closing airports, and controlling prices of goods.	587
7. CURE	News about good and bad cure, diagnosis, ventilators, supportive medical equipment, etc.	116
8. VOLUNT	Volunteering efforts or donation of money, goods or services.	133
9. PRSNL	Personal story or opinion.	453
10. SUPPORT	Support or praise governments, medical staff, celebrities, etc.	386
11. PRAYER	Prayer.	563
12. UNIMP	Unrelated or unimportant such as spams or advertisements.	786
13. NOT_ARB	Not Arabic, e.g. Persian.	161
Total		**8000**

Table 1: Annotation classes and distribution: Important classes (top) and LessImportant classes (bottom)

Figure 1: Examples for REP, ACT, INFO, RUMOR, ADVICE and SEEK_ACT classes

Figure 2: Examples for CURE, VOLUNT, PRSNL, SUPPORT, PRAYER and UNIMP classes

Table 2 shows country distribution and top accounts for the original authors of tweets.

Typically, people retweet tweets from ministry of health in their countries in addition to famous news agencies and celebrities. Most of these accounts are verified.

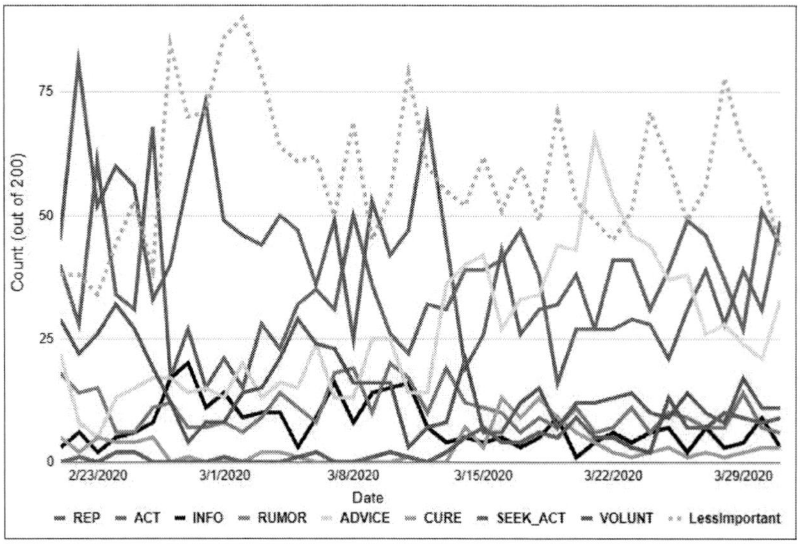

Figure 3: Timeline: Number of class tweets each day

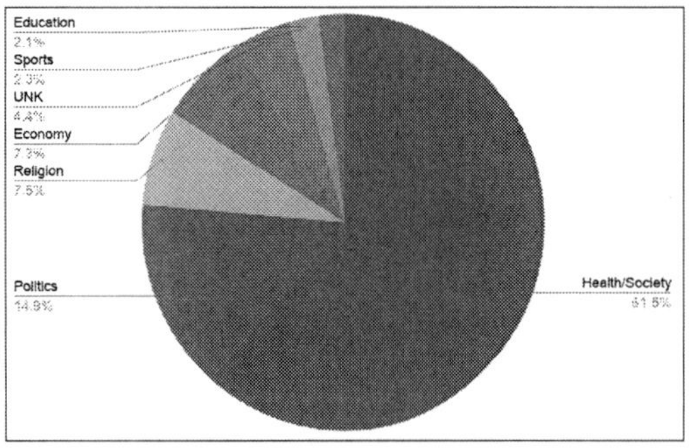

Figure 4: Tweet topics

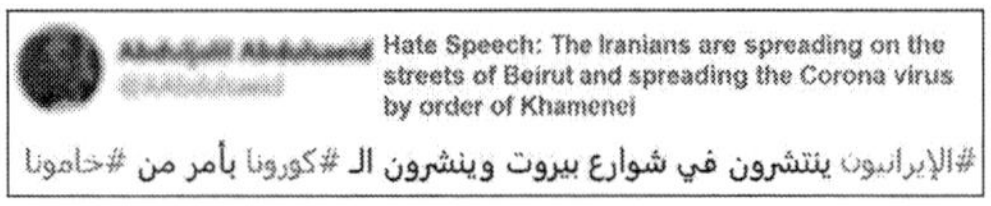

Figure 5: Hate speech example

Country	%	Top Accounts
SA	59	SaudiNews50, SaudiMOH
OTH	13	MohamadAhwaze, amjadt25
OM	7	OmanVSCovid19, OmaniMOH
KW	7	Almajlliss, KUWAIT_MOH
QA	4	amansouraja, MOPHQatar
AE	4	AlHadath, AlArabiya_Brk
EG	3	RassdNewsN, mohpegypt

Table 2: Country distribution and top accounts

6 Experiments and Evaluation

We randomly split the data into sets of 6000, 1000 and 1000 tweets for train, dev and test sets respectively. We report macro-averaged Precision (P), Recall (R) and F1 score along with Accuracy (Acc) on test set [6]. We use F1 score as primary metric for comparison.

6.1 Features

N-gram features We experimented with character and word n-gram features weighted by term frequency-inverse term document frequency (tf-idf). We report results for only the most significant ranges, namely, word [1-2] and character [2-5].

Mazajak Embeddings Mazajak embeddings are word-level skip-gram embeddings trained on 250M Arabic tweets, yielding 300-dimensional vectors (Abu Farha and Magdy, 2019).

6.2 Classification Models

Support Vector Machines (SVMs) SVMs have been shown to perform decently for Arabic text classification tasks such as spam detection (Mubarak et al., 2020), adult content detection (Mubarak et al., 2021; Hassan et al., 2021), offensiveness detection (Hassan et al., 2020b,a) or dialect identification (Abdelali et al., 2020; Bouamor et al., 2019). We experimented with i) word n-gram, ii) character n-gram and iii) Mazajak Embeddings. We used LinearSVC implementation by scikit-learn [7].

[6]Differences between dev and test sets are $\pm 2-3\%$ (F1).

[7]https://scikit-learn.org/

Model	Features	Acc.	P	R	F1
Majority Class	-	72.5	36.3	50.0	42.0
SVM	W[1-2]	84.4	82.5	76.4	78.6
SVM	C[2-5]	**85.4**	**84.3**	77.4	**79.8**
SVM	Mazajak	83.9	80.5	77.7	78.9
AraBERT		83.9	80.0	**79.2**	79.6

Table 3: Binary classification results

Model	Features	Acc.	P	R	F1
Majority Class	-	22.7	1.7	7.7	2.8
SVM	W[1-2]	**62.8**	**64.3**	53.5	56.3
SVM	C[2-5]	59.0	**64.3**	49.4	51.8
SVM	Mazajak	60.0	55.1	51.5	52.4
AraBERT		62.7	61.6	**59.8**	**60.5**

Table 4: Fine-grained classification results

Deep Contextualized Transformer Models (BERT) Transformer-based pre-trained contextual embeddings, such as BERT (Devlin et al., 2019), have outperformed other classifiers in many NLP tasks. We used AraBERT (Antoun et al., 2020), a BERT-based model trained on Arabic news. We used ktrain library (Maiya, 2020) that utilizes Huggingface[8] implementation to fine-tune AraBERT. We used learning rate of $8e^{-5}$, truncating length of 50 and fine-tuned for 5 epochs.

6.3 Binary Classification

First, we experiment to distinguish LessImportant tweets from others (see Section 4). From Table 3, we can see that SVMs with character [2-5]-gram outperformed others with F1 score of **79.8**, closely followed by AraBERT with **79.6** F1.

6.4 Fine-grained Classification

Our next set of experiments were designed for fine-grained classification for 13 classes. With F1 score of **60.5**, AraBERT outperformed others (Table 4).

Error Analysis: AraBERT confusion matrix (Figure 6) shows that PRSNL, INFO and RUMOR are the hardest classes to identify and the most common error is misclassifying INFO as ADVICE. We hypothesize these errors can be reduced if larger data set is being annotated.

7 Conclusion and Future Work

We present the largest publicly available manually annotated dataset of Arabic tweets for 13 classes that includes the most retweeted tweets in the early days of COVID-19. Followed by data analysis, we present models that can reliably identify important tweets and can perform fine-grained classification. In the future, we plan to compare our data to data from later days of the pandemic.

[8]https://huggingface.co/

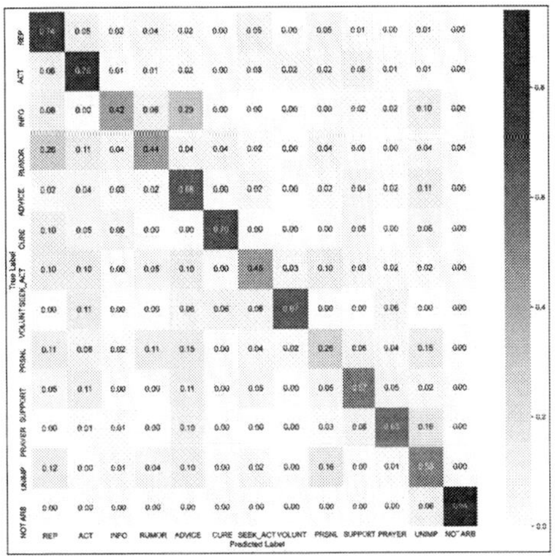

Figure 6: Confusion matrix normalized over true labels

References

Ahmed Abdelali, Hamdy Mubarak, Younes Samih, Sabit Hassan, and Kareem Darwish. 2020. Arabic dialect identification in the wild. *ArXiv*, abs/2005.06557.

Ibrahim Abu Farha and Walid Magdy. 2019. Mazajak: An online Arabic sentiment analyser. In *Proceedings of the Fourth Arabic Natural Language Processing Workshop*, pages 192–198, Florence, Italy. Association for Computational Linguistics.

Firoj Alam, Shaden Shaar, Fahim Dalvi, Hassan Sajjad, Alex Nikolov, Hamdy Mubarak, Giovanni Da San Martino, Ahmed Abdelali, Nadir Durrani, Kareem Darwish, and Preslav Nakov. 2020. Fighting the covid-19 infodemic: Modeling the perspective of journalists, fact-checkers, social media platforms, policy makers, and the society. *arXiv:2005.00033*.

Sarah Alqurashi, Ahmad Alhindi, and Eisa Alanazi. 2020. Large arabic twitter dataset on covid-19. *arXiv preprint arXiv:2004.04315*.

Thayer Alshaabi, David R. Dewhurst, Joshua R. Minot, Michael V. Arnold, Jane L. Adams, Christopher M. Danforth, and Peter Sheridan Dodds. 2020. The growing amplification of social media: Measuring temporal and social contagion dynamics for over 150 languages on twitter for 2009–2020. Available online at http://arxiv.org/abs/2003.03667.

Lama Alsudias and Paul Rayson. 2020. COVID-19 and Arabic Twitter: How can Arab world governments and public health organizations learn from social media? In *Proceedings of the 1st Workshop on NLP for COVID-19 at ACL 2020*, Online. Association for Computational Linguistics.

Wissam Antoun, Fady Baly, and Hazem M. Hajj. 2020. Arabert: Transformer-based model for arabic language understanding. *ArXiv*, abs/2003.00104.

Juan M Banda, Ramya Tekumalla, Guanyu Wang, Jingyuan Yu, Tuo Liu, Yuning Ding, and Gerardo Chowell. 2020. A large-scale covid-19 twitter chatter dataset for open scientific research–an international collaboration. *arXiv preprint arXiv:2004.03688*.

Houda Bouamor, Sabit Hassan, and Nizar Habash. 2019. The MADAR shared task on Arabic fine-grained dialect identification. In *Proceedings of the Fourth Arabic Natural Language Processing Workshop*, pages 199–207, Florence, Italy. Association for Computational Linguistics.

Emily Chen, Kristina Lerman, and Emilio Ferrara. 2020. Tracking social media discourse about the covid-19 pandemic: Development of a public coronavirus twitter data set. *JMIR Public Health Surveill*, 6(2):e19273.

Matteo Cinelli, Walter Quattrociocchi, Alessandro Galeazzi, Carlo Michele Valensise, Emanuele Brugnoli, Ana Lucia Schmidt, Paola Zola, Fabiana Zollo, and Antonio Scala. 2020. The covid-19 social media infodemic. *arXiv preprint arXiv:2003.05004*.

Jacob Devlin, Ming-Wei Chang, Kenton Lee, and Kristina Toutanova. 2019. BERT: Pre-training of deep bidirectional transformers for language understanding. In *Proceedings of the 2019 Conference of the North American Chapter of the Association for Computational Linguistics: Human Language Technologies, Volume 1 (Long and Short Papers)*, pages 4171–4186, Minneapolis, Minnesota. Association for Computational Linguistics.

Fatima Haouari, Maram Hasanain, Reem Suwaileh, and Tamer Elsayed. 2020. ArCOV-19: The first Arabic COVID-19 twitter dataset with propagation networks. *arXiv:2004.05861*.

Sabit Hassan, Hamdy Mubarak, Ahmed Abdelali, and Kareem Darwish. 2021. Asad: Arabic social media analytics and understanding. In *Proceedings of the Software Demonstrations of the 16th Conference of the European Chapter of the Association for Computational Linguistics*.

Sabit Hassan, Younes Samih, Hamdy Mubarak, and Ahmed Abdelali. 2020a. ALT at SemEval-2020 task 12: Arabic and English offensive language identification in social media. In *Proceedings of the Fourteenth Workshop on Semantic Evaluation*, pages 1891–1897, Barcelona (online). International Committee for Computational Linguistics.

Sabit Hassan, Younes Samih, Hamdy Mubarak, Ahmed Abdelali, Ammar Rashed, and Shammur Absar Chowdhury. 2020b. ALT submission for OSACT shared task on offensive language detection. In *Proceedings of the 4th Workshop on Open-Source Arabic Corpora and Processing Tools, with a Shared Task on Offensive Language Detection*, pages 61–65, Marseille, France. European Language Resource Association.

Binxuan Huang and Kathleen M Carley. 2020. Disinformation and misinformation on twitter during the novel coronavirus outbreak. *arXiv preprint arXiv:2006.04278*.

Yan Leng, Yujia Zhai, Shaojing Sun, Yifei Wu, Jordan Selzer, Sharon Strover, Julia Fensel, Alex Pentland, and Ying Ding. 2020. Analysis of misinformation during the covid-19 outbreak in china: cultural, social and political entanglements. *arXiv preprint arXiv:2005.10414*.

L. Li, Q. Zhang, X. Wang, J. Zhang, T. Wang, T. Gao, W. Duan, K. K. Tsoi, and F. Wang. 2020. Characterizing the propagation of situational information in social media during covid-19 epidemic: A case study on weibo. *IEEE Transactions on Computational Social Systems*, 7(2):556–562.

Arun S Maiya. 2020. ktrain: A low-code library for augmented machine learning. *arXiv preprint arXiv:2004.10703*.

Hamdy Mubarak, Ahmed Abdelali, Sabit Hassan, and Kareem Darwish. 2020. Spam detection on arabic twitter. In *Social Informatics*, pages 237–251, Cham. Springer International Publishing.

Hamdy Mubarak, Sabit Hassan, and Ahmed Abdelali. 2021. Adult content detection on arabic twitter: Analysis and experiments. In *Proceedings of the Sixth Arabic Natural Language Processing Workshop*.

Umair Qazi, Muhammad Imran, and Ferda Ofli. 2020. Geocov19. *SIGSPATIAL Special*, 12(1):6–15.

Gautam Kishore Shahi, Anne Dirkson, and Tim A Majchrzak. 2020. An exploratory study of covid-19 misinformation on twitter. *arXiv preprint arXiv:2005.05710*.

Qiang Yang, Hind Alamro, Somayah Albaradei, Adil Salhi, Xiaoting Lv, Changsheng Ma, Manal Alshehri, Inji Jaber, Faroug Tifratene, Wei Wang, et al. 2020. Senwave: Monitoring the global sentiments under the covid-19 pandemic. *arXiv preprint arXiv:2006.10842*.

Multilingual Negation Scope Resolution for Clinical Text

Mareike Hartmann
Department of Computer Science
University of Copenhagen
Denmark
hartmann@di.ku.dk

Anders Søgaard
Department of Computer Science
University of Copenhagen
Denmark
soegaard@di.ku.dk

Abstract

Negation scope resolution is key to high-quality information extraction from clinical texts, but so far, efforts to make encoders used for information extraction negation-aware have been limited to English. We present a universal approach to multilingual negation scope resolution, that overcomes the lack of training data by relying on disparate resources in different languages and domains. We evaluate two approaches to learn from these resources, training on combined data and training in a multi-task learning setup. Our experiments show that zero-shot scope resolution in clinical text is possible, and that combining available resources can improve performance in most cases.

1 Introduction

Information extraction (IE) from clinical text, such as electronic health records (EHR) or clinical trial narratives, is a promising application of machine learning that can potentially benefit many areas of the health sector (Dalianis, 2018). IE systems are applied to facilitate administrative tasks by assigning medical codes (Stanfill et al., 2010), to extract phenotype information about patients (Gehrmann et al., 2018), and to improve patient care by monitoring healthcare associated infections (Proux et al., 2011) and adverse drug events (ADE) (Luo et al., 2017). Results are promising, but the majority of work focuses on clinical text in English (Névéol et al., 2018). This puts at a disadvantage patients in countries where health narratives are recorded in languages other than English. This gap can be overcome by multilingual IE systems that are applicable to text in multiple languages. Such multilingual systems can, on the one hand, lead towards improving healthcare within non-English speaking countries. On the other hand, they allow to gather information across countries, which is particularly interesting for rare diseases with few cases per country, and can increase the statistical power of an analysis (Jensen et al., 2012; Névéol et al., 2018). With this paper, we contribute to improving IE from clinical text in languages other than English, which is a step towards improving healthcare for all.

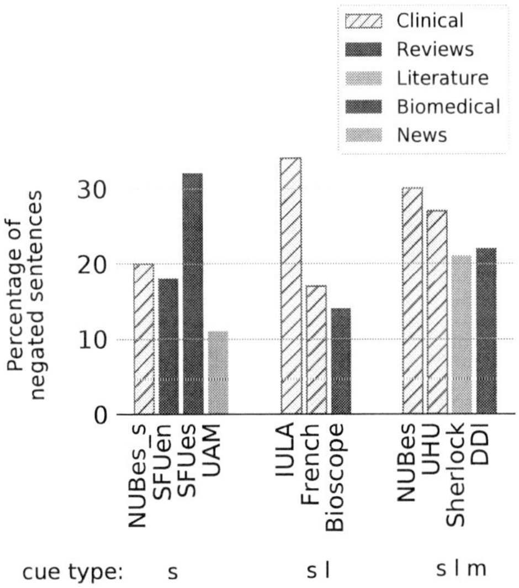

Figure 1: Percentage of negated sentences in clinical and non-clinical text. Datasets are grouped according to the type of negation cue they annotate (syntactic (s), lexical (l), morphological (m), see further details in). Numbers for datasets not listed in Section 4 (UAM (Sandoval and Salazar, 2013) and UHU (Diaz et al., 2017)) are taken from (Jiménez-Zafra et al., 2020). NUBes_s counts only syntactic cues in NUBES.

Negation is a phenomenon that has received considerable attention in IE models for clinical and biomedical text, as there is large interest in identifying negated concepts (Mutalik et al., 2001; Chapman et al., 2001; Mowery et al., 2012), e.g. negated medical events (Nawaz et al., 2013) or negated drug-drug interactions (Bokharaeian et al., 2016). Absence of symptoms or the fact that chemical re-

Proceedings of the 12th International Workshop on Health Text Mining and Information Analysis, pages 7–18
April 19, 2021. ©2021 Association for Computational Linguistics

actions are not observable are crucial knowledge in the clinical and biomedical domain (Elkin et al., 2005; Krallinger, 2010). This is reflected in Figure 1, which shows the percentage of negated sentences in clinical compared to non-clinical text. Explicitly integrating negation information into machine learning (ML) models can improve performance for relation extraction (Chowdhury and Lavelli, 2013) and more general NLP tasks such as sentiment analysis (Barnes et al., 2020) and machine translation (Fancellu and Webber, 2014), which become increasingly popular for the clinical domain (Denecke and Deng, 2015). Even though pre-trained language models tailored to the biomedical and clinical domains (Lee et al., 2020; Peng et al., 2019; Alsentzer et al., 2019) now produce state-of-the-art results for several downstream tasks, such language models do not sufficiently capture the semantics of negation (Ettinger, 2020; Kassner and Schütze, 2020).

One way of accessing negation information is to explicitly detect all negated words in a sentence, a task which is referred to as *negation scope resolution* (Morante et al., 2008). The task has recently been successfully approached by fine-tuning a pre-trained language model on labeled target data (Sergeeva et al., 2019; Khandelwal and Sawant, 2020). For the clinical domain however, we cannot rely on multilingual labeled target data to be available. This is partly due to data privacy issues, which lead to few publicly available datasets in the clinical domain (Chapman et al., 2011; Velupillai et al., 2018) [1], and partly due to the fact that other languages are underrepresented compared to English in clinical NLP (Névéol et al., 2018).[2] In summary, when building a multilingual negation scope resolution system for clinical text, we are facing a lack of training data. Our approach is hence to use the available data as best as possible to build a negation resolution model that works on data in multiple languages.

Negation scope resolution has also been considered for non-clinical text (Morante and Sporleder, 2012; Morante and Blanco, 2012), and several datasets spanning various domains and including a small amount of non-English languages are available (see Section 4). In this work, we investigate if and how these disparate data sources can serve as training resource to resolve negation scope in multiple languages in the clinical domain in a zero-shot setup. To enable transfer across languages we rely on multilingual BERT (mBERT) (Devlin et al., 2019), a multilingual pre-trained language encoder that has proven capable of zero-shot cross-lingual transfer in other tasks[3] (Wu and Dredze, 2019), and has recently been applied for zero-shot transfer to French clinical text (Shaitarova et al., 2020). One challenge arising from the available scope resolution datasets is that they are annotated according to different annotation schemes (see Section 3.3), raising the question if and how they can be combined as a training resource (Barnes et al., 2020; Jiménez-Zafra et al., 2020).

We explore two strategies to handle disparities between the resources, a simple concatenation of datasets after a partial conversion of annotations, and a multi-task learning (MTL) setup, where each dataset is handled as a different task. The MTL setup also allows us to explore additional auxiliary tasks that can potentially help in resolving negation scope by making use of available resources. Here, we consider a classification task for negated events in the clinical domain.

Contributions We study an approach for zero-shot cross-lingual transfer for negation scope resolution in clinical text, exploiting data from disparate sources by data concatenation, or in an MTL setup.[4] We demonstrate that it is possible to achieve decent performance across Spanish and French clinical texts without scope annotated training data in the target language, or in the target domain.[5] We further explore an auxiliary task that makes use of negated medical event detection data, however finding that they cannot improve performance.

2 Related Work

Recent work used large pre-trained language models, in particular BERT, for negation scope resolution as a sequence labeling task (Sergeeva et al., 2019; Khandelwal and Sawant, 2020) and produced

[1]For example, there are several clinical datasets of EHRs annotated with negation, but they are not publicly available (see Chapter 4.7.3 in Dalianis (2018)).

[2]This trend can also be observed in NLP datasets outside of the clinical domain (Bender, 2011). Also in domains other than the clinical domain, the range of languages in publicly available datasets for negation scope resolution is small (Jiménez-Zafra et al., 2020).

[3]https://sites.research.google/xtreme

[4]We make our code and trained models available at https://github.com/coastalcph/multi_neg_scope

[5]We however rely on a small annotated validation dataset in target language and domain for choosing the best model.

(1) El ojo derecho **no** completa la aducción .

(2) **No** existen datos de focalidad neurológica **ni** signos de meningismo.

(3) Es **incapaz de** levantarse de la silla **sin** ayuda

Table 1: Example sentences from the IULA corpus. Negation cues are marked in bold, negation scopes are underlined and the overbrace indicates additional annotations of medical concepts.

new state-of-the-art results on the BIOSCOPE and SHERLOCK datasets. Barnes et al. (2020) deviate from the two-step sequence labeling approach and propose a sequence labeling model that detects cues and scope in one step. Kurtz et al. (2020) show that it is beneficial to frame the negation scope resolution task as a graph parsing problem. All the works listed above focus exclusively on English data, and only few works attempt to do multi- or cross-lingual negation scope resolution.

Fancellu et al. (2018) were the first to present a zero-shot approach for negation scope resolution. They find that transfer from English to a Chinese version of the SHERLOCK corpus is possible, using an LSTM and a graph convolutional network (GCN) in combination with static cross-lingual word embeddings. However, both their models rely on PoS-tags and more importantly on dependency parses of the input sentences.[6]

Shaitarova et al. (2020) present zero and few-shot experiments for the FRENCH clinical data and Spanish review data using mBERT fine-tuned on a concatenation of English datasets. Their model is equivalent to our single task model trained on concatenated data ($\mathbf{ST}_{\text{cat}}$ in Section 6). While their work demonstrates the general applicability of the approach, we focus on the applicability to clinical text in multiple languages, by exploring more training resources, and more importantly by evaluating the generalization performance of the approach across three clinical datasets in two languages.

3 Negation Scope Resolution

Negation is a phenomenon in language that changes the truth value of a proposition. From a linguistic perspective, a negation signal or cue, i.e. an expression that indicates negation, is an operator, which has a scope over parts of a sentence.[7]

In example sentence (1) in Table 1 above, the negation cue **no** (denoted in bold) affects the underlined parts of the sentence, which are referred to as the scope of this negation cue. In many clinical IE tasks, the goal is to resolve the negation of medical concepts or entities, i.e. to determine if they are present or absent, for example the procedure of eye movement in (1) and the medical finding of focal signs and signs of meningism in (2). This can be approached as a binary classification task predicting if the medical entity is negated or not (Chapman et al., 2001). Depending on the application, however, one might want to resolve negation of a different set of medical concepts (such as relations), in which case the classifier has to be re-trained on the new set of concepts. Morante et al. (2008) suggest to first find the whole scope of the negation cue [8], and then check if the concepts of interest are contained in the scope or not. With this setup, a system can be re-used across concepts, and does not have to be re-trained for a different downstream task. Also, as negation is inherent in text from any domain, this approach opens the possibility to learn to resolve negation scope in other domains, and transfer this knowledge to clinical text.

3.1 Resolving Negation Scope

The negation scope resolution task is usually solved in two steps: First, negation cues are identified, often using a lexicon. Second, the scope of these negation cues, i.e. the words that are affected by the negation cue, are identified. The second step is approached as a sequence labeling task, where given a negation cue, each word in the sequence is labeled with respect to whether it is affected by this cue (in scope) or not (out of scope). In the second

[6]Obtaining dependency parses for text from several domains and in several languages requires expensive adaptation of parsers, and is not applicable for our setup. Hence, Fancellu et al. (2018) is not included as a baseline in our experiments.

[7]In this work, we only consider intra-sentential negation, i.e. negation that affects words within the sentence containing the negation cue.

[8]This setup is now commonly referred to as negation scope resolution (Morante and Blanco, 2012).

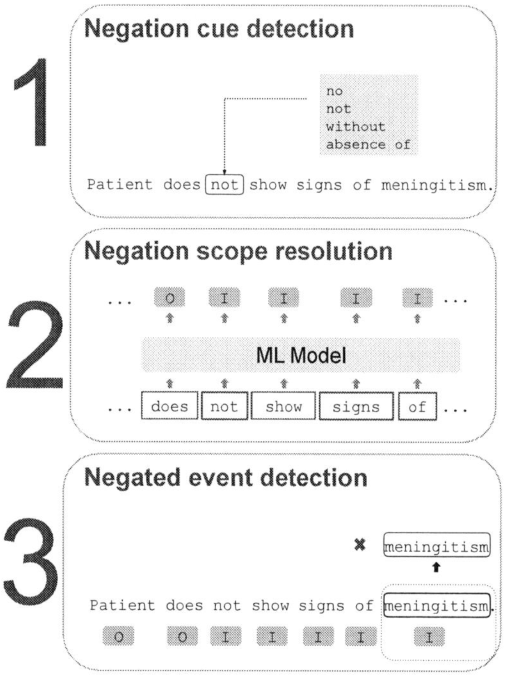

Figure 2: Pipeline for resolving negation in clinical data. First, negation cues are identified using a list of pre-defined negation cues. Second, the scope of the detected negation cue is identified using a machine learning model, which assigns a binary scope label (either I for in- or O for out-of-scope) to each token in the sequence. Finally, the predicted negation scope can be used to identify negated events. Our work focuses on the second step. As ML model, we use an MTL architecture with a shared pre-trained multilingual encoder.

step, information about the cue is provided as input to the model, and the model handles one cue per sequence, i.e. sentences with multiple negation cues are represented using multiple input sequences with one marked cue each.[9] Figure 2 shows the individual steps of this pipeline approach.

3.2 Cue Detection

Most recent work, including ours, focuses on the second step of the pipeline and resolves negation scope given gold cues, arguing that negation cues in a practical setting can best be identified using a lexicon-based approach. We think this is a reasonable approach, as in the annotation process for many datasets, cue annotation involves a pre-defined lexicon. Figure 3 shows that a fixed set of negation cues accounts for the majority of negations in the data. Following Jiménez-Zafra et al. (2020), we discriminate between three types of negation cues. Syntactic negation cues such as **not** and **without** form the largest group of cues. Lexical negation cues are words with a meaning that indicates a negation, such as **lack of**. Morphological negation cues are words that contain a morpheme expressing the negation, such as prefix **a-** in **a**symptomatic. We do not include unsupervised experiments for cue detection, as the available datasets annotate different types of cues (see Table 2).

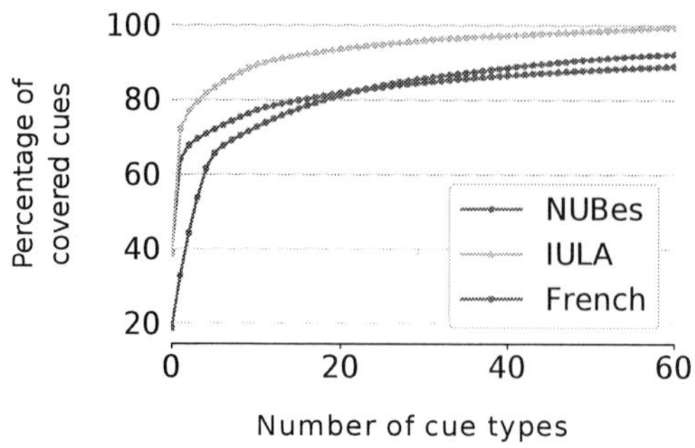

Figure 3: Distribution of negation cues in the datasets. A small number of different cues is responsible for a large amount of negations.

3.3 Difference in Annotation Schemes

In addition to annotating different types of negation cues, the available datasets also follow different annotation schemes for elements to include in the scopes of these cues, e.g. if subjects, cue tokens and punctuation are included in the scope or not. In Table 2, we list annotation features of the datasets involved. In their review on corpora annotated with negation, Jiménez-Zafra et al. (2020) emphasize these differences in annotation and state that *as a consequence, it would not be possible to merge all of them to training a negation processing system.* Problems with differences in annotation are also noted by Barnes et al. (2020), who find that concatenating datasets for training does not lead to improved results. We propose two steps to alleviate this problem: If possible, we convert different annotations to a similar scheme. In particular, we

[9]Barnes et al. (2020) propose a model that does cue detection and scope resolution in one pass, by adding a special cue label for cues in the scope resolution labeling task. Here, the scope is not conditioned on a specific cue, which means with several cues in a sentence, we do not know which cue is associated with a predicted scope. While this setup seems more practical to use with a downstream task in mind, the two stage setup allows to have a dedicated language specific cue detection component, which is particular interesting in a cross-lingual zero-shot setup.

Dataset	Statistics: Domain	Lang	# Train	Annotation: Type	Subj	Cue	Punc
IULA	clinical records	es	771	l s	~	×	✓
NUBES	clinical records	es	5,297	l s m	~	×	×
FRENCH	clinical records	fr	1,272	l s	✓	×	×
BIOSCOPE	biomedical publications	en	1,469	l s	×	✓	×
SHERLOCK	short stories	en	618	l s m	✓	×	×
SFUES	reviews	es	2,796	s	~	✓	×
SFUEN	reviews	en	2,458	s	×	×	×
DDI	biomedical database	en	892	l s m	?	✓	✓

Table 2: Dataset statistics and difference in annotation schemes. The test split sizes of the IULA, NUBES, and FRENCH datasets are 173, 1152 , and 272 sentences, respectively. Negation types are either lexical (l), syntactic (s), or morphological (m). ~ signifies that the subject is included in the scope only in specific cases. Features in the last two columns can easily be converted to a standardized annotation (include cue token, exclude punctuation). Negation type and inclusion/exclusion of subject cannot be accounted for without extra effort.

always include cue tokens in the scope, and exclude punctuation if the punctuation marker denotes the end of a scope. The other differences in annotation, in particular inclusion of subjects and differences in annotated cue type, cannot be equalized easily. Here, we aim at still learning as much as possible from the available data with different annotation schemes by using an MTL model with output layers specific for each dataset, and hence tailored to each annotation scheme.

4 Datasets

In the following, we describe the clinical datasets as well as the training resources used in our experiments. A more detailed overview of annotation guidelines, negation types and negation components can be found in the survey of Jiménez-Zafra et al. (2020), as well as in the original works associated with the datasets. Statistics on the available data can be found in Table 2.

4.1 Clinical Negation Scope Resolution Datasets

IULA The IULA corpus (Marimon et al., 2017) is a collection of Spanish clinical records from *several services of one of the main hospitals in Barcelona (Spain)* (Marimon et al., 2017) and includes text from five sections of the electronic record: physical exploration, evolution, radiology, current process, and comparative explorations. Here, syntactic and lexical negation cues are annotated. Subjects are (almost) always excluded from the scope if they precede the verb. The corpus also contains annotations for four types of medical entities: body structure, substance, clinical finding, and procedure.

NUBes The NUBES corpus (Lopez et al., 2020) is a collection of anonymized Spanish clinical records from a private hospital. The text is extracted from seven sections present in the electronic record: chief complaint, present illness, physical examination, diagnostic tests, surgical history, progress notes, therapeutic recommendations. Here, syntactic, lexical, and morphological negation cues are annotated.

French The FRENCH corpus (Dalloux et al., 2019) is a collection of clinical trial protocols in French which were obtained from the registry of the Gustave Roussy hospital as well as the French National Cancer Institute. The corpus includes text parts about the patient inclusion criteria and the description of the procedure of the trials. Here, syntactic and lexical negation cues are annotated. In this corpus, there is no marked association between a negation cue and its scope, hence sentences with different negation cues have to be processed at once.

4.2 Other Negation Scope Resolution Datasets

Bioscope The BIOSCOPE corpus (Vincze et al., 2008) is a collection of English biomedical and clinical texts and consists of three parts: abstracts and full papers of biomedical publications, and radiology reports. Unfortunately, we were not able to obtain access to the radiology reports, hence in our experiments we only include the biomedical publications (abstracts and full papers).

Sherlock The SHERLOCK corpus (Morante and Daelemans, 2012) comprises two English Conan Doyle short stories. It was used in the 2012 Shared Task on negation scope resolution and is the most popular benchmark corpus. Here, syntactic, lexical and morphological negation cues are annotated.

DDI The DDI corpus (Bokharaeian et al., 2014) is a collection of texts from the chemical and pharmaceutical database DrugBank, and English Medline abstracts.

SFU$_{en}$ The SFUEN corpus (Konstantinova et al., 2012) is a collection of product reviews in English. The reviews are extracted from the reviewing website *Epinions.com* and cover 8 different categories[10].

SFU$_{es}$ The SFUES corpus (Jiménez-Zafra et al., 2018) is a collection of Spanish reviews from the reviewing website *Ciao.es* also covering 8 different categories.

4.3 Auxiliary Task

One obvious choice of auxiliary task is the detection of negated events, which is a sequence classification task, where the event is gold annotated and replaced by a special token.[11] We use the instances labeled as *present* and *absent* in the English M2C2 assertion dataset (Uzuner et al., 2011) in order to generate a binary sentence classification task.

5 Approach

MTL Setup The easiest way to join the disparate training resources annotated for negation scope resolution is to simply concatenate them (Barnes et al., 2020; Shaitarova et al., 2020). However, the difference between annotations in the negation scope resolution datasets (see Section 3.3) makes it reasonable to use an MTL setup that aims at learning several tasks at once in order to better solve a target task. Here, we treat each training dataset as a separate task. The MTL setup also allows to exploit negation information in datasets associated with tasks other than negation scope resolution, e.g. negated event detection. We use an MTL model with hard parameter sharing (Caruana, 1997; Collobert et al., 2011), i.e. all model parameters except for the task specific output layers are shared between the different tasks. The model is shown in Figure 4. Using multilingual BERT as shared multilingual encoder enables the model to learn from datasets in different languages, and to do zero-shot cross-lingual transfer at inference time. In particular, we run our experiments using the MT-DNN framework[12] (Liu et al., 2019).

Sampling During training, we sample batches from a task i according to $p_i \propto N_i$, where N_i is the size of the dataset associated with task i.We implement annealed sampling following Stickland and Murray (2019) as sampling according to $p_i \propto N_i^{\alpha}$, with $\alpha = 1 - f\frac{e-1}{E-1}$. Here, e refers to the current epoch, E is the maximum number of training epochs and f a fixed annealing value.[13] The idea is that α decreases while training progresses, and hence the difference in dataset size becomes less prevalent later in the training. In the last epoch, batches are sampled from each task with equal probability, regardless of dataset size. For all experiments, we fix the number of total batches sampled in each epoch to the number of distinct batches over all datasets. We experiment with two variants for sampling training instances. We either only train on sentences that contain a negation, or we also include sentences that do not contain any negation. In the latter setting, we downsample the amount of sentences without negation to be equal to the number of negated sentence (which means that in this setting, the models are trained on twice as much data as in the negated-only setting). Both sampling variants are treated as hyperparameters and we select the best setting on the validation data.

Zero-shot Setup We assume no training data for the clinical datasets and test in a zero-shot setup, after converting all annotations according to the steps described in Section 3.3. During training, we use per-token F1-scores on the development split of the BIOSCOPE corpus for model selection. To identify the combination of training datasets that are expected to perform best on the target datasets,

[10] books, cars, computers, cookware, hotels, films, music, and phones

[11] Difference and similarities between the tasks of negated event detection and negation scope resolution are discussed in (Stenetorp et al., 2012)

[12] `https://github.com/namisan/mt-dnn`

[13] Following Stickland and Murray (2019) we set $f = 0.8$. We set E to the number of training epochs.

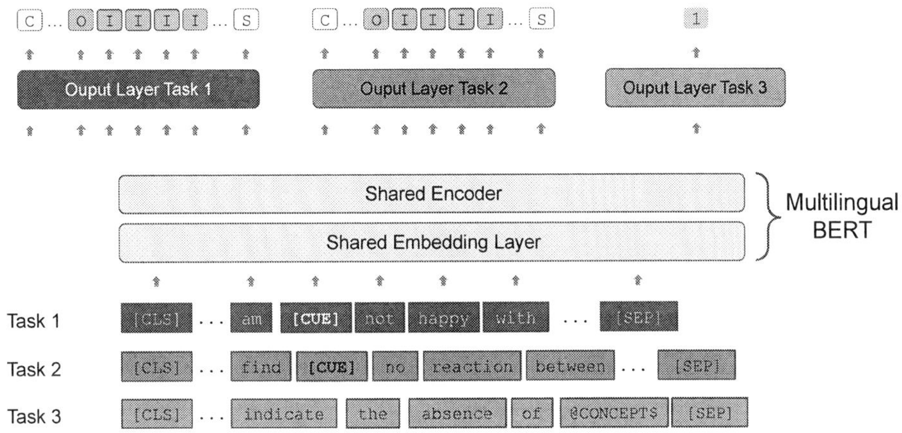

Figure 4: MTL model with a shared multilingual encoder for multilingual negation scope resolution. The multilingual embedding and encoder layers are shared between all the tasks, whereas each task has a task-specific output layer. Task 1 and 2 are sequence labeling tasks for negation scope resolution on product reviews and biomedical text, respectively. Task 3 is a binary sequence classification task on the M2C2 dataset. For the negation scope resolution tasks, information about the cue is added to the sequence in the form of a special token preceding the cue token.

we use F1-scores averaged over the development splits of the three clinical datasets. This reflects the use-case of a user annotating a small set of target sentences, in order to choose among the best pre-trained negation scope resolvers that can be readily applied to the specific target dataset.

5.1 Pretrained Multilingual Encoder

As pre-trained multilingual encoder, we use multilingual BERT, a transformer-based language model pre-trained with a masked language modeling and next sentence prediction objective on Wikipedia text in multiple languages.[14] The encoder can easily be adapted to downstream classification tasks by putting a randomly initialized classification layer on top of the pre-trained encoder.[15] mBERT has shown cross-lingual transfer abilities when fine-tuned on target data as well as in zero-shot setups (Wu and Dredze, 2019; Artetxe et al., 2020).

Following the standard input scheme (Devlin et al., 2019), for all fine-tuning tasks, we add a special `[CLS]` token serving as an aggregated sentence representation to the beginning of each sentence, and a special `[SEP]` token indicating the end of a sequence to the end of the sentence.

Sequence Labeling For the sequence labeling tasks, we assign labels for each sub-token. Negation cue information is provided to the model by inserting a special `[CUE]` token before the negation cue (see Figure 4).[16] Special labels are assigned to the `[CUE]` token, `[CLS]` token, `[SEP]` token and subword tokens that are not the first subword of a word. These are ignored during evaluation, which means for evaluation a word is considered correctly labeled if the first subword token generated from that word receives the correct label.

Sequence Classification For the sequence classification task, we use the `[CLS]` token as aggregated sentence representation and assign a label to the whole sequence by feeding the encoded `[CLS]` token representation through the output classification layer.

[14] We use the `bert-base-multilingual-cased` version.

[15] In our experiments, we fine-tune both the encoder and the classification layer on the downstream task.

[16] Khandelwal and Sawant (2020) found this method to work better than *replacing* the negation cue by the special token. Sergeeva et al. (2019) suggest to add a special cue embedding to the subtoken embedding at the input layer, which we found to perform comparable to the insertion method on the BIOSCOPE abstracts.

	Model	Resources	Test data					
			IULA		NUBES		FRENCH	
			F1	PCS	F1	PCS	F1	PCS
H	Punct	-	91.09	84.38	79.37	68.06	85.32	45.96
SU	mBERT	in-domain	$96.23_{0.23}$	$89.38_{0.63}$	$95.66_{0.22}$	$88.02_{0.78}$	$95.01_{0.28}$	$81.49_{0.77}$
ZS	**ST**	SFUEN	$93.73_{0.12}$	$88.13_{0.63}$	$89.42_{0.50}$	$75.35_{2.02}$	$88.63_{0.11}$	$55.27_{1.53}$
	ST$_{\text{cat}}$	BIOSCOPE + SFUEN	$94.21_{0.88}$	$87.29_{1.44}$	$90.24_{0.21}$	$77.08_{1.31}$	$89.15_{0.90}$	$52.34_{4.07}$
	MTL$_{\text{n}}$	BIOSCOPE, SFUEN	$94.24_{1.52}$	$86.67_{4.16}$	$89.41_{0.34}$	$75.67_{2.16}$	$87.93_{0.89}$	$47.43_{5.12}$
	MTL$_{\text{n+e}}$	SFUEN, M2C2	$93.49_{0.57}$	$85.63_{1.65}$	$89.88_{0.41}$	$74.33_{2.33}$	$87.89_{0.23}$	$53.29_{1.39}$

Table 3: Results for negation scope resolution with gold cues on the clinical datasets (with converted annotations). Scores are averaged over three runs with different random seeds, with standard deviations as subscripts. We report results for heuristics (**H**), supervised (**SU**) and zero-shot (**ZS**) experiments. For ZS, we report the best single task model (**ST**) with training on concatenated data (**ST$_{\text{cat}}$**), and multi-task models trained on negation scope resolution tasks (**MTL$_{\text{n}}$**) or with and additional event detection task (**MTL$_{\text{n+e}}$**). Best performance in the **ZS** setup is marked in gray.

5.2 Evaluation

We report two widely used evaluation metrics for negation scope prediction: Percentage of correct spans (PCS) and F1 over scope tokens (Morante and Blanco, 2012). The latter is the standard F1-score computed on the token level, whereas PCS is computed on the span level and considers a predicted span correct if it exactly matches the gold span. PCS is stricter in general and due to different annotation schemes (with respect to inclusion of subject and type of negation cue), PCS is a challenging criterion in zero-shot setups when models are trained on an annotation scheme that differs from the target annotation scheme.

6 Experiments

We split all datasets into 70% for training, 15% for validation, 15% for testing, and predict negation scope in the test splits of the three clinical datasets IULA, NUBES, and FRENCH. Results are shown in Table 3.[17] We report zero-shot performances (**ZS**) of a single task model trained on a single dataset (**ST**) and on a concatenation of datasets (**ST$_{\text{cat}}$**). For the multi-task model, we either train on several negation scope resolution tasks (**MTL$_{\text{n}}$**), or add an additional event detection task (**MTL$_{\text{n+e}}$**). For each model, we report the best combination of training datasets determined by the best average F1-score across the validation splits of the three clinical datasets. Validation scores for other dataset combinations can be found in our code repository.

As baselines, we report in-domain performance of mBERT , which provides an upper baseline indicating the performance gain if annotated target data was available. We also include a punctuation baseline, as previous work has found that many test sentences are easy to label from cue token to next punctuation mark (Fancellu et al., 2017; Sergeeva et al., 2019). The punctuation baseline labels all tokens following the negation cue until the next punctuation mark[18] as *in scope*. In all experiments, mBERT is fine-tuned with the default hyperparameters for sequence labeling tasks implemented in MT-DNN, which means that we use the adamax optimizer with a learning rate of 5e-5 and a batch size of 8. The maximum sequence length is set to 512.[19] All models are trained for a maximum of 50 epochs, and the best model is selected according to F1-score on the BIOSCOPE validation split.

6.1 Results

Baselines The in-domain experiments show that NUBES and FRENCH are harder to predict than IULA. The performance of the punctuation baseline as measured by F1-score is already high, but

[17]Note that our results are reported on the test sets with converted annotations, and hence not directly comparable to numbers reported on the original data.

[18]We consider as punctuation marks any of the following characters: .,:;!?()[]

[19]Sequences longer than the maximum sequence length are truncated, rather than split into shorter sequences. However, this affects none of the sequence in our test sets.

still considerably lower than the performance of the in-domain models.

Zero-shot For the zero-shot experiments, we report the best **ST** model trained on a single dataset across all test sets and metrics, which is the one trained on SFUEN. This is surprising, because neither domain, nor language, nor annotated cue types in SFUEN correspond to the features of the test data.

Combining resources by concatenating datasets improves performance according to F1-score across all datasets. $\mathbf{ST}_{\text{cat}}$ outperforms $\mathbf{MTL}_{\text{n}}$ in all scenarios, except for F1 on IULA where they perform comparably. Hence, even though training datasets are annotated with different annotation schemes that cannot easily be converted to a common scheme, simply concatenating the datasets and training a single-task model seems to be more effective than using a multi-task model.

In addition, we find that adding event detection as an auxiliary task in the $\mathbf{MTL}_{\text{n+e}}$ model cannot improve over the models that only do scope resolution, and in some cases are even detrimental. Overall, we see that it is possible to resolve negation scope in clinical text without labeled training data, and that the a single-task model trained on concatenated data works best.

7 Conclusion

Negation is a frequent phenomenon in clinical text, and resolving its scope can benefit clinical IE tasks. As clinical IE has huge potential to improve healthcare, it should ideally be available regardless of language. Availability in multiple languages, however, is hindered by a lack of annotated data for fine-tuning language models in task-specific data in the target language. In this paper, we show that even without labeled data in the target domain or target language, negation scope resolution in clinical text is possible, by fine-tuning a multilingual language model on available resources from other domains and languages. Even though annotation schemes for negation scope differ, combining available resources for training improves performance in most cases.

Acknowledgements

The authors thank the Lundbeck foundation for funding their research in the context of the Brain-Drugs project.

References

Emily Alsentzer, John Murphy, William Boag, Wei-Hung Weng, Di Jindi, Tristan Naumann, and Matthew McDermott. 2019. Publicly available clinical bert embeddings. In *Proceedings of the 2nd Clinical Natural Language Processing Workshop*, pages 72–78.

Mikel Artetxe, Sebastian Ruder, and Dani Yogatama. 2020. On the cross-lingual transferability of monolingual representations. In *Proceedings of the 58th Annual Meeting of the Association for Computational Linguistics*, pages 4623–4637, Online. Association for Computational Linguistics.

Jeremy Barnes, Erik Velldal, and Lilja Øvrelid. 2020. Improving sentiment analysis with multi-task learning of negation. *Natural Language Engineering*, pages 1–21.

Emily M Bender. 2011. On achieving and evaluating language-independence in nlp. *Linguistic Issues in Language Technology*, 6(3):1–26.

Behrouz Bokharaeian, Alberto Diaz, and Hamidreza Chitsaz. 2016. Enhancing extraction of drug-drug interaction from literature using neutral candidates, negation, and clause dependency. *PloS one*, 11(10).

Behrouz Bokharaeian, Alberto Diaz, Mariana Neves, and Virginia Francisco. 2014. Exploring negation annotations in the drugddi corpus. In *Fourth workshop on building and evaluating resources for health and biomedical text processing (BIOTxtM 2014)*, pages 1–8. Citeseer.

Rich Caruana. 1997. Multitask learning. *Machine learning*, 28(1):41–75.

Wendy W Chapman, Will Bridewell, Paul Hanbury, Gregory F Cooper, and Bruce G Buchanan. 2001. A simple algorithm for identifying negated findings and diseases in discharge summaries. *Journal of biomedical informatics*, 34(5):301–310.

Wendy W Chapman, Prakash M Nadkarni, Lynette Hirschman, Leonard W D'Avolio, Guergana K Savova, and Ozlem Uzuner. 2011. Overcoming barriers to NLP for clinical text: the role of shared tasks and the need for additional creative solutions. *Journal of the American Medical Informatics Association*, 18(5):540–543.

Md Faisal Mahbub Chowdhury and Alberto Lavelli. 2013. Exploiting the scope of negations and heterogeneous features for relation extraction: A case study for drug-drug interaction extraction. In *Proceedings of the 2013 Conference of the North American Chapter of the Association for Computational Linguistics: Human Language Technologies*, pages 765–771.

Ronan Collobert, Jason Weston, Léon Bottou, Michael Karlen, Koray Kavukcuoglu, and Pavel Kuksa. 2011. Natural language processing (almost) from scratch.

Journal of Machine Learning Research, 12:2493–2537.

Hercules Dalianis. 2018. *Clinical text mining: Secondary use of electronic patient records*. Springer Nature.

Clément Dalloux, Natalia Grabar, and Vincent Claveau. 2019. Détection de la négation: corpus français et apprentissage supervisé. *Revue des Sciences et Technologies de l'Information-Série TSI: Technique et Science Informatiques*, pages 1–21.

Kerstin Denecke and Yihan Deng. 2015. Sentiment analysis in medical settings: New opportunities and challenges. *Artificial intelligence in medicine*, 64(1):17–27.

Jacob Devlin, Ming-Wei Chang, Kenton Lee, and Kristina Toutanova. 2019. Bert: Pre-training of deep bidirectional transformers for language understanding. In *Proceedings of the 2019 Conference of the North American Chapter of the Association for Computational Linguistics: Human Language Technologies, Volume 1 (Long and Short Papers)*, pages 4171–4186.

Noa P Cruz Diaz, Roser Morante, Manuel J Mana López, Jacinto Mata Vázquez, and Carlos L Parra Calderón. 2017. Annotating negation in spanish clinical texts. In *Proceedings of the workshop computational semantics beyond events and roles*, pages 53–58.

Peter L Elkin, Steven H Brown, Brent A Bauer, Casey S Husser, William Carruth, Larry R Bergstrom, and Dietlind L Wahner-Roedler. 2005. A controlled trial of automated classification of negation from clinical notes. *BMC medical informatics and decision making*, 5(1):13.

Allyson Ettinger. 2020. What BERT is not: Lessons from a New Suite of Psycholinguistic Diagnostics for Language Models. *Transactions of the Association for Computational Linguistics*, 8:34–48.

Federico Fancellu, Adam Lopez, and Bonnie Webber. 2018. Neural networks for cross-lingual negation scope detection. *arXiv preprint arXiv:1810.02156*.

Federico Fancellu, Adam Lopez, Bonnie Webber, and Hangfeng He. 2017. Detecting negation scope is easy, except when it isn't. In *Proceedings of the 15th Conference of the European Chapter of the Association for Computational Linguistics: Volume 2, Short Papers*, pages 58–63.

Federico Fancellu and Bonnie Webber. 2014. Applying the semantics of negation to smt through n-best list re-ranking. In *Proceedings of the 14th Conference of the European Chapter of the Association for Computational Linguistics*, pages 598–606.

Sebastian Gehrmann, Franck Dernoncourt, Yeran Li, Eric T Carlson, Joy T Wu, Jonathan Welt, John Foote Jr, Edward T Moseley, David W Grant, Patrick D Tyler, et al. 2018. Comparing deep learning and concept extraction based methods for patient phenotyping from clinical narratives. *PloS one*, 13(2):e0192360.

Peter B Jensen, Lars J Jensen, and Søren Brunak. 2012. Mining electronic health records: towards better research applications and clinical care. *Nature Reviews Genetics*, 13(6):395–405.

Salud María Jiménez-Zafra, Roser Morante, María Teresa Martín-Valdivia, and L Alfonso Ureña-López. 2020. Corpora annotated with negation: An overview. *Computational Linguistics*, 46(1):1–52.

Salud María Jiménez-Zafra, Mariona Taulé, M Teresa Martín-Valdivia, L Alfonso Ureña-López, and M Antònia Martí. 2018. Sfu review sp-neg: a spanish corpus annotated with negation for sentiment analysis. a typology of negation patterns. *Language Resources and Evaluation*, 52(2):533–569.

Nora Kassner and Hinrich Schütze. 2020. Negated and Misprimed Probes for Pretrained Language Models: Birds Can Talk, But Cannot Fly. In *Proceedings of the 58th Annual Meeting of the Association for Computational Linguistics*, pages 7811–7818.

Aditya Khandelwal and Suraj Sawant. 2020. NegBERT: A transfer learning approach for negation detection and scope resolution. In *Proceedings of The 12th Language Resources and Evaluation Conference*, pages 5739–5748, Marseille, France. European Language Resources Association.

Natalia Konstantinova, Sheila CM De Sousa, Noa P Cruz Díaz, Manuel J Mana López, Maite Taboada, and Ruslan Mitkov. 2012. A review corpus annotated for negation, speculation and their scope. In *Lrec*, pages 3190–3195.

Martin Krallinger. 2010. Importance of negations and experimental qualifiers in biomedical literature. In *Proceedings of the Workshop on Negation and Speculation in Natural Language Processing*, pages 46–49.

Robin Kurtz, Stephan Oepen, and Marco Kuhlmann. 2020. End-to-end negation resolution as graph parsing. In *Proceedings of the 16th International Conference on Parsing Technologies and the IWPT 2020 Shared Task on Parsing into Enhanced Universal Dependencies*, pages 14–24, Online. Association for Computational Linguistics.

Jinhyuk Lee, Wonjin Yoon, Sungdong Kim, Donghyeon Kim, Sunkyu Kim, Chan Ho So, and Jaewoo Kang. 2020. Biobert: a pre-trained biomedical language representation model for biomedical text mining. *Bioinformatics*, 36(4):1234–1240.

Xiaodong Liu, Pengcheng He, Weizhu Chen, and Jianfeng Gao. 2019. Multi-task deep neural networks for natural language understanding. In *Proceedings of ACL*, pages 4487–4496, Florence, Italy. Association for Computational Linguistics.

Salvador Lima Lopez, Naiara Perez, Montse Cuadros, and German Rigau. 2020. Nubes: A corpus of negation and uncertainty in spanish clinical texts. In *Proceedings of The 12th Language Resources and Evaluation Conference*, pages 5772–5781.

Yuan Luo, William K Thompson, Timothy M Herr, Zexian Zeng, Mark A Berendsen, Siddhartha R Jonnalagadda, Matthew B Carson, and Justin Starren. 2017. Natural language processing for ehr-based pharmacovigilance: a structured review. *Drug safety*, 40(11):1075–1089.

Montserrat Marimon, Jorge Vivaldi, and Núria Bel. 2017. Annotation of negation in the IULA Spanish clinical record corpus. In *Proceedings of the Workshop Computational Semantics Beyond Events and Roles*, pages 43–52, Valencia, Spain. Association for Computational Linguistics.

Roser Morante and Eduardo Blanco. 2012. * sem 2012 shared task: Resolving the scope and focus of negation. In ** SEM 2012: The First Joint Conference on Lexical and Computational Semantics–Volume 1: Proceedings of the main conference and the shared task, and Volume 2: Proceedings of the Sixth International Workshop on Semantic Evaluation (SemEval 2012)*, pages 265–274.

Roser Morante and Walter Daelemans. 2012. Conandoyle-neg: Annotation of negation cues and their scope in conan doyle stories. In *Proceedings of the Eighth International Conference on Language Resources and Evaluation (LREC'12)*, pages 1563–1568.

Roser Morante, Anthony Liekens, and Walter Daelemans. 2008. Learning the scope of negation in biomedical texts. In *Proceedings of the 2008 conference on empirical methods in natural language processing*, pages 715–724.

Roser Morante and Caroline Sporleder. 2012. Modality and negation: An introduction to the special issue. *Computational linguistics*, 38(2):223–260.

Danielle L Mowery, Sumithra Velupillai, and Wendy Chapman. 2012. Medical diagnosis lost in translation–Analysis of uncertainty and negation expressions in English and Swedish clinical texts. In *BioNLP: Proceedings of the 2012 Workshop on Biomedical Natural Language Processing*, pages 56–64.

Pradeep G Mutalik, Aniruddha Deshpande, and Prakash M Nadkarni. 2001. Use of general-purpose negation detection to augment concept indexing of medical documents: a quantitative study using the umls. *Journal of the American Medical Informatics Association*, 8(6):598–609.

Raheel Nawaz, Paul Thompson, and Sophia Ananiadou. 2013. Negated bio-events: analysis and identification. *BMC bioinformatics*, 14(1):14.

Aurélie Névéol, Hercules Dalianis, Sumithra Velupillai, Guergana Savova, and Pierre Zweigenbaum. 2018. Clinical natural language processing in languages other than english: opportunities and challenges. *Journal of biomedical semantics*, 9(1):12.

Yifan Peng, Shankai Yan, and Zhiyong Lu. 2019. Transfer learning in biomedical natural language processing: An evaluation of bert and elmo on ten benchmarking datasets. In *Proceedings of the 18th BioNLP Workshop and Shared Task*, pages 58–65.

Denys Proux, Caroline Hagège, Quentin Gicquel, Suzanne Pereira, Stefan Darmoni, Frédérique Segond, and Marie Hélène Metzger. 2011. Architecture and systems for monitoring hospital acquired infections inside hospital information workflows. In *Proceedings of the Second Workshop on Biomedical Natural Language Processing*, pages 43–48.

Antonio Moreno Sandoval and Marta Garrote Salazar. 2013. La anotación de la negación en un corpus escrito etiquetado sintácticamente. *Revista Iberoamericana de Linguistica*, 8:45–61.

Elena Sergeeva, Henghui Zhu, Amir Tahmasebi, and Peter Szolovits. 2019. Neural Token Representations and Negation and Speculation Scope Detection in Biomedical and General Domain Text. In *Proceedings of LOUHI 2019*, pages 178–187, Hong Kong. Association for Computational Linguistics.

Anastassia Shaitarova, Lenz Furrer, and Fabio Rinaldi. 2020. Cross-lingual transfer-learning approach to negation scope resolution. In *Swiss Text Analytics Conference & Conference on Natural Language Processing 2020*.

Mary H Stanfill, Margaret Williams, Susan H Fenton, Robert A Jenders, and William R Hersh. 2010. A systematic literature review of automated clinical coding and classification systems. *Journal of the American Medical Informatics Association*, 17(6):646–651.

Pontus Stenetorp, Sampo Pyysalo, Tomoko Ohta, Sophia Ananiadou, and Jun'ichi Tsujii. 2012. Bridging the gap between scope-based and event-based negation/speculation annotations: A bridge not too far. In *Proceedings of the Workshop on Extra-Propositional Aspects of Meaning in Computational Linguistics*, pages 47–56, Jeju, Republic of Korea. Association for Computational Linguistics.

Asa Cooper Stickland and Iain Murray. 2019. Bert and pals: Projected attention layers for efficient adaptation in multi-task learning. In *International Conference on Machine Learning*, pages 5986–5995.

Özlem Uzuner, Brett R South, Shuying Shen, and Scott L DuVall. 2011. 2010 i2b2/va challenge on concepts, assertions, and relations in clinical text. *Journal of the American Medical Informatics Association*, 18(5):552–556.

Sumithra Velupillai, Hanna Suominen, Maria Liakata, Angus Roberts, Anoop D Shah, Katherine Morley, David Osborn, Joseph Hayes, Robert Stewart, Johnny Downs, et al. 2018. Using clinical natural language processing for health outcomes research: Overview and actionable suggestions for future advances. *Journal of biomedical informatics*, 88:11–19.

Veronika Vincze, György Szarvas, Richárd Farkas, György Móra, and János Csirik. 2008. The bioscope corpus: biomedical texts annotated for uncertainty, negation and their scopes. *BMC bioinformatics*, 9(11):1–9.

Shijie Wu and Mark Dredze. 2019. Beto, bentz, becas: The surprising cross-lingual effectiveness of bert. In *Proceedings of the 2019 Conference on Empirical Methods in Natural Language Processing and the 9th International Joint Conference on Natural Language Processing (EMNLP-IJCNLP)*, pages 833–844.

Understanding Social Support Expressed in a COVID-19 Online Forum

Anietie Andy[1], Brian Chu [1], Ramie Fathy [1],
Barrington Bennet [1], Daniel Stokes [1], Sharath Chandra Guntuku [2]
[1]Penn Medicine [2]Computer and Information Science
University of Pennsylvania
{anietie.andy, Brian.Chu, Ramie.Fathy,
Barrington.Bennett ,Daniel.Stokes}
@pennmedicine.upenn.edu
sharathg@cis.upenn.edu

Abstract

In online forums focused on health and well-being, individuals tend to seek and give the following social support: emotional and informational support. Understanding the expressions of these social supports in an online COVID-19 forum is important for: (a) the forum and its members to provide the right type of support to individuals and (b) determining the long term effects of the COVID-19 pandemic on the well-being of the public, thereby informing interventions. In this work, we build four machine learning models to measure the extent of the following social supports expressed in each post in a COVID-19 online forum: (a) emotional support given (b) emotional support sought (c) informational support given, and (d) informational support sought. Using these models, we aim to: (i) determine if there is a correlation between the different social supports expressed in posts e.g. when members of the forum give emotional support in posts, do they also tend to give or seek informational support in the same post? (ii) determine how these social supports sought and given changes over time in published posts. We find that (i) there is a positive correlation between the informational support given in posts and the emotional support given and emotional support sought, respectively, in these posts and (ii) over time, users tended to seek more emotional support and give less emotional support.

1 Introduction

Globally, millions of individuals have contracted COVID-19 and more than 2 million people have died from the pandemic as of January 2021 [1]. Individuals are turning to online forums focused on discussions around COVID-19 to seek and give support (Stokes et al., 2020). In online health and well-being forums, individuals tend to seek and give two forms of social support: emotional and informational support (Wang et al., 2012; Yang et al., 2017); where: (a) emotional support sought seeks understanding, affirmation and encouragement, (b) emotional support given includes providing encouragement, (c) informational support sought seeks advice or information, and (d) informational support given provides advice and information. Below are examples (rephrased) of posts that express these social supports in a COVID-19 related online forum:

- **Emotional Support Given:** "I wish everyone on this forum good health however if any of us tests positive for COVID please keep us informed on how you are doing. We as a community will offer emotional support and also learn how it feels to have the virus"
- **Emotional Support Sought:** "Will we be forced to stay home and not work for some weeks? I don't think I can afford 2 weeks without getting payed."
- **Informational Support Given:** "A 14-day quarantine makes sense. This will ensure that you are not going to become symptomatic and prevent pre-symptomatic spread."
- **Informational Support Sought:** "When a vaccine is found, would countries be required to pay for a license to produce the vaccine?"

In this work, given posts published on an online COVID-19 forum, we build 4 machine learning models to measure the extent of the following social supports: (i) emotional support given, (ii) emotional support sought, (iii) informational support given, and (iv) informational support sought.

In online forums, some posts may express multiple social supports; hence, understanding if there is a correlation between multiple social supports expressed in posts is important for better understanding the social support needs of members of the forum. We conduct an analysis to determine

[1]https://coronavirus.jhu.edu/map.html

Proceedings of the 12th International Workshop on Health Text Mining and Information Analysis, pages 19–27
April 19, 2021.

the correlation between different social supports expressed in posts published in the COVID-19 online forum.

These social supports sought or given may change over time. Understanding how these social support needs change over time in a COVID-19 online forum is important for several reasons; first, it can help the forum moderators and members provide adequate and meaningful support to users, and secondly, it can provide insights about the long term effects of the pandemic on the well-being of the public, thereby providing the necessary authorities with sufficient and timely information on the changes in social support needs of the public so that they can make resources available to address these needs. We carry out an analysis to determine the gaps in social support supply and demand in posts published on an online COVID-19 forum over time. We conduct additional analysis to identify different time periods in which there are clear differences in the types of social support sought or given and determine language feature differences in posts published in these different time periods.

2 Related Work

A lot of prior work has been done to gain insights and respond to the COVID-19 pandemic.

In order to identify treatments and gain insights about COVID-19, the Allen Institute for AI in partnership with the White House and other collaborators, created a resource to use various artificial intelligence and natural language techniques to access thousands of research papers and articles related to COVID-19 to retrieve pertinent information about published research work about COVID-19 (Wang et al., 2020). Using this resource, several research work has been done: (i) in (Esteva et al., 2020), an automated system was built in which given a query, it retrieves relevant publications related to the query and ranks them, (ii) in (Verspoor et al., 2020), a system was developed to help medical researchers easily find research articles relevant to COVID-19, and (iii) in (Hope et al., 2020), a system was built which identifies the relationships between researchers working on COVID-19 related projects and their networks with the aim to help researchers interested in COVID-19 related projects to easily identify potential collaborators and research ideas to investigate. In (Wei et al., 2020), a dataset of more than 1,500 questions from several sources were obtained and annotated to different categories and a model was built to classify questions into these categories.

A resource made up of a dashboard was created by researchers at Johns Hopkins University for tracking the number of cases, information about contact tracing, testing, and information about work on the vaccine for COVID-19 [2]. In (Guntuku et al., 2020), using Twitter data, a dashboard was created to track (over time) the changes in language use (related to mental health and COVID-19 symptoms), in different states in the United States. Similarly, the World Health Organisation created a dashboard for monitoring the number of COVID-19 cases globally[3].

Some prior work used social media data to gain insights into peoples reactions to the COVID-19 pandemic. Posts from a Reddit forum focused on discussions around COVID-19 were analyzed to understand the patterns in topics discussed and how they change over time in the forum (Stokes et al., 2020). In Zhang et al. (2020), two Reddit forums focused on discussions around COVID-19 were analyzed to gain insights about peoples responses to the pandemic. In Zong et al. (2020), Twitter data related to COVID-19 events such as positive and negative COVID-19 tests were labelled and models were developed to automatically identify these events. In (Chen et al., 2020), a multilingual COVID-19 Twitter dataset was created with the aim to aid researchers better understand what people are conversing about on social media, as it relates to the pandemic and enable researchers study COVID-19 related misinformation. In (Ayers et al., 2020), internet searches during the time period of the pandemic were studied and it was determined that peoples expression of anxiety increased. In (Serrano et al., 2020), a method was created for detecting COVID-19 related misinformation in Youtube videos by using the comments to the videos. In (Aggarwal et al., 2020), the differences in language use in COVID-19 related Reddit discussions between female and male users was studied.

The main difference between our work and prior work is that in this work we build 4 machine learning models to measure different social supports expressed in posts in an online COVID-19 forum with the aim to better understand the types of social support people were expressing in these posts and

[2] `https://coronavirus.jhu.edu/map.html`
[3] `https://covid19.who.int/`

how they changed over time.

3 Dataset

Our dataset consists of posts published on a Reddit forum, */r/Coronavirus* - which is focused on discussions around COVID-19 and is the COVID-19 subreddit with the most number of members (i.e. 2.4 million members as of January 2021). Specifically, we collected 64,074 posts published daily in the "Daily Discussion Post" thread in the */r/Coronavirus* subreddit between March 3 and April 30 2020 (Stokes et al., 2020). Table 1 shows information about the number of posts published in March and April, respectively. Figure 1 shows information about the number of posts published weekly in our dataset; it can be observed that the number of posts published weekly peaked in March and steadily declined afterwards.

Table 1: Number of posts in our dataset published in March and April 2020

Month	Number of Posts
March 2020	45,880
April 2020	18,194

Figure 1: Graph showing the number of posts published weekly in our dataset

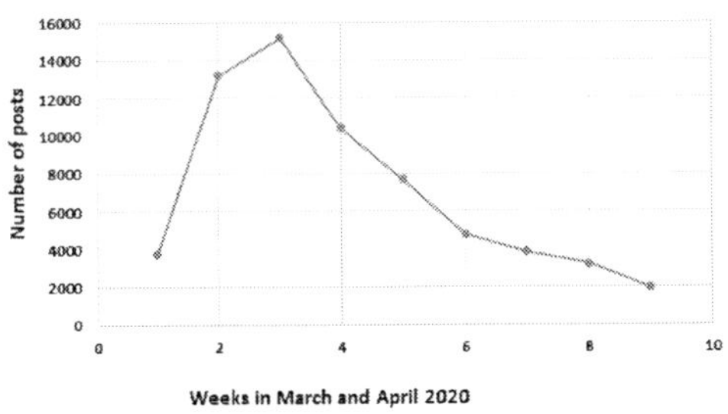

In this work, we aim to determine the types of social support being expressed in posts published in the */r/Coronavirus* forum; while analyzing the comment responses to these posts would give insights as to how users respond to posts in this forum, our goal in this work is to better understand the social support expressed in posts in this forum and how it changes over time. Hence, in our analysis, we exclude the comments these posts received.

4 Social Support

The importance of social support expressed in online health forums has been demonstrated (Wang et al., 2012; Yang et al., 2017).

Similar to prior work (Wang et al., 2012), we measure the extent of social support expressed in posts. We had 3 medical students rate a randomly selected sample of 1,000 posts from our dataset; Table 2 shows a brief description of the sample posts.

Table 2: Summary of sample dataset

Attributes	Result
Avg. number of sentences per post	3.6
Avg. number of words per sentence	52.53
Avg. standard deviation of number of sentences	2.23
Avg. standard deviation of words per sentence	39.5

Given a post, the annotators rated the posts on: (i) the extent of emotional support given, (ii) the extent of emotional support sought, (iii) the extent of informational support given, and (iv) the extent of informational support sought. Specifically, using a 7-point Likert scale, where 1 meant "a particular social support was not expressed" and 7 meant "a particular social support was expressed a lot" the annotators rated the 1,000 sample posts. For each post, the annotators ratings were aggregated by averaging their ratings and each post had a numeric score between 1 and 7 expressing the extent of emotional support given, emotional support sought, informational support given, and informational support sought, respectively in the post. Using intra-class correlation (ICC) (Bartko, 1966), we measured the reliability of the annotators; the ICC for emotional support given, emotional support sought, informational support given, and informational support sought was 0.650, 0.660, 0.778, 0.918, respectively.

We extracted the following features from each post (Wang et al., 2012):

4.1 Features

We extract the following Linguistic Inquiry and Word Count (LIWC) (Pennebaker et al., 2015) categories which are relevant to emotional and informational support: "cognitive mechanism", "verbs", "religion", "negative emotion", "positive emotion", "biological processes", "present focus", "past focus", "future focus", "time", "death" , impersonal pronoun", personal pronouns (i.e. "1st person singular", "1st person plural", "2nd person", "3rd person singular", "3rd person plural"). LIWC is a psycholinguistic dictionary made up of several categories with curated words associated with each category (Wang et al., 2012).

From each post we: (1) count the number of

words and sentences, (2) count the number of sentences that contain negation words such as "not", (3) count the number sentences which are phrased as questions; these sentences were identified as follows: (i) sentences that ended with a question mark, (ii) sentences that began with a modal verb e.g. "could you", and (iii) sentences that began with an indirect question pattern e.g. "would it be possible" (Wang et al., 2012).

The number of certain parts-of-speech tags, strong subjectivity words e.g. "abuse" and weak subjectivity words e.g. "absence" (Wilson et al., 2005) were counted (Wang et al., 2012).

We identified and counted posts involving requests or advice (Wang et al., 2012) by using the following word patterns: (a) *"you"* and a modal verb, which counts the number of sentences that start with the pronoun *"you"* and then a modal verb e.g. "might" and (b) *"please"* and a modal verb which identify sentences that start with the word *"please"* and then a verb (Wang et al., 2012).

Using the Mallet (McCallum, 2002) implementation of Latent Dirichlet Allocation (LDA) (Blei et al., 2003), we generated 20 topics from posts in our dataset (section 3) (we varied the number of topics and 20 topics gave the most coherent topic themes when examined by one of the co-authors) (Wang et al., 2012); Table 3 shows the top 5 words associated with the LDA topics.

4.2 Social Support Prediction Model

Given the language features extracted from posts in section 4.1, using Scikit learn (Pedregosa et al., 2011), we built four machine learning Random Forest regression models. For each post, these models produced numerical values (between 1 and 7) indicating the extent of: (i) emotional support given, (ii) emotional support sought, (iii) informational support given, and (iv) informational support sought. The 1,000 annotated posts from section 4 were randomly partitioned into training set (80%), validation (10%), and testing set (10%).The validation set was used to evaluate the models performance and when a satisfactory performance was obtained, the models were then evaluated on the testing set. To measure the models performance, we used Pearson's correlation (Wang et al., 2012) to correlate the models outputs with the average annotator labels for each post.

Table 3: Top 5 words associated with LDA topics

Highly correlated words associated with LDA topics
work, home, job, company, pay
symptom, fever, cough,sick, flu
people, think, saying,right, pandemic
good, essential, thanks, great, hope
lockdown, China, virus, million, infected
care, hospital, medical, health, workers
family, quarantine, self, friend, parents
people, risk, high, immune, old
virus, hands, spread, wash, infected
days, weeks, today, ago, yesterday
positive, test, tested, gloves, negative
question, corona, ask, answer, curve
coronavirus, covid, flu, vaccine, sars
state, people, city, died, county
death, rate, social, distancing, recovered
home, food, store, going, grocery
open, closed, online, school, America
mask, use, wear, face, protect
cases, deaths, numbers, testing, Italy
months, long, end, states, travel

4.3 Result from Social Support Prediction Model

On the annotated dataset (section 4), these models correlated with the average annotator ratings as shown in Table 4. The models were then applied to all the posts in our dataset (section 3). Each post ended up having scores for the emotional support given, emotional support sought, informational support given, and informational support sought, respectively. Table 5 and Table 6 shows the 5 most most important features (these features are ranked by the Random Forest models).

Table 4: Results from the analysis to determine the correlation between the social support models and the annotator ratings: all the p-values were less that 0.001

Social Support	Pearson's Correlation
Emotional Support Given	0.400
Emotional Support Sought	0.468
Informational Support Given	0.503
Informational Support Sought	0.547

Table 5: Top 5 most important features as ranked by the Random forest model for emotional support given and emotional support sought

Emo. Support Given	Emo. Support Sought
Number of words (0.188)	Number of words (0.213)
Strong Subjectivity (0.165)	Noun (0.171)
Noun (0.145)	Strong subjectivity (0.147)
Adjective (0.126)	Adjective (0.111)
Weak subjectivity(0.109)	Sentence length (0.101)

Table 6: Top 5 most important features as ranked by the Random forest model for informational support given and informational support sought

Info. Support Given	Info. Support Sought
Number of words (0.196)	Question (0.199)
Noun (0.173)	Number of words (0.169)
Strong subjectivity (0.115)	Noun (0.136)
Sentence length (0.091)	Strong subjectivity (0.112)
Question (0.072)	Sentence length (0.098)

5 Correlate the different types of Social Supports expressed in posts

In this section, we aim to determine if there are correlations between the types of social supports expressed in posts in our dataset. This is important for better understanding the support needs of the individuals publishing posts on the forum, and can guide the moderators of the forum and users responding to these posts to tailor their responses accordingly to match the support needs expressed in the post.

For each post in our dataset (section 3), given features that represent the predicted scores for the emotional support given, emotional support sought, informational support given, and informational support sought, respectively, we use the Pearson's correlation to correlate the feature that represents the emotional support given with the features that represent the informational support given and informational support sought, respectively. Similarly, we correlate the feature that represents the emotional support sought with the features that represents the informational support given and the informational support sought, respectively. Tables 7 and 8 show these correlations.

Table 7: Correlation between the emotional support given and the informational support given and sought, respectively. The p-values are less than 0.05

	Emo. Support Given
Informational Support Given	Pearson r = 0.721
Informational Support Sought	Pearson r = -0.069

Table 8: Correlation between the emotional support sought and the informational support given and sought, respectively. The p-values are less than 0.05

	Emo. Support Sought
Informational Support Given	Pearson r = 0.381
Informational Support Sought	Pearson r = 0.019

6 Changes in Social Support expressed in posts Over Time

In this section, we aim to determine how social supports expressed in posts in our dataset changes over time. Given the posts in our dataset and features that represent their corresponding scores for emotional support given, emotional support sought, informational support given, and informational support sought, respectively, we aim to determine the statistical differences between these features and then determine how these social supports change over time.

First, for all the posts in our dataset, we use t-test to determine the difference between the features that represent the emotional support given and emotional support sought, respectively; also, we use t-test to determine the difference between the features that represents the informational support given and informational support sought, respectively, in posts in our dataset. The results from the t-tests are shown in Table 9.

Table 9: The results of the t-tests. The p-values were less than 0.05. *Emo. Given* indicates the emotional support given, *Emo. Sought* indicates the emotional support sought, *Info. Given* indicates the informational support given, and *Info. Sought* indicates the informational support sought

	T-value
Emo. Given and Emo. Sought	- 51.56
Info. Given and Info. Sought	176.97

Second, we aim to determine how these social supports changed over the time period in which we collected our dataset i.e March and April 2020. For posts published in the same week, we averaged their scores for emotional support given, emotional support sought, informational support given, and informational support sought, respectively. As shown in Figure 2, we observed that emotional support sought increased over time and the emotional support given increased in the beginning of March and reduced by April. Also, we observed that informational support given and sought followed a similar pattern, however, users published more posts in which they gave informational support, as shown in Figure 3.

7 Language features associated with time periods with clear differences in the types of social supports expressed in posts

From section 6, it was observed that the emotional support sought increased over time and the emotional support given increased in March and then decreased over time.

Figure 2: Graph showing the change in the average weekly emotional support sought and emotional support given in posts from our /r/Coronavirus dataset

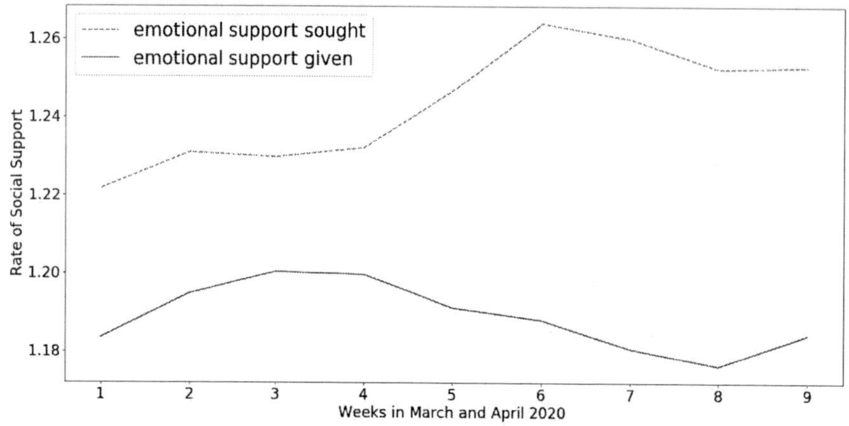

Table 10: Topics associated with posts from our dataset published in March

LDA Topics and highly correlated words	Cohen's D
travel, school, week, trip, cancel, flight, break, online,university, college	0.183
home, family, live, stay, parents, house, mom, sick, dad, friend	0.109
symptoms, fever, cough, days, throat, sick, flu, feel, cold, coughing	0.102

In this section, we aim to determine the language features associated with the time periods in which there was a clear change in the emotional support sought and emotional support given. Specifically, from figure 2, we determine week 4 (which corresponds to the end of March 2020) as the start of the ascent of the emotional support sought and the descent of the emotional support given, hence using the following language features: LDA (Blei et al., 2003) and LIWC (Pennebaker et al., 2015), we aim to determine the LDA topics and LIWC categories associated with posts published in these different time periods (i.e. March 03 - March 31 and April 01 - April 30 2020). Potentially, the LDA topics and LIWC categories associated with these different time periods should give some insights as to why there was a change in the emotional support given and emotional support sought.

8 LDA

We identified topics associated with posts published between March 03 2020 - March 31 2020 and April 01 - April 30 2020 in our datast by generating 20 LDA topics (Blei et al., 2003) from posts in our dataset; we tried different numbers of topics and 20 topics obtained the most coherent topic themes when examined by one of the co-authors. Tables 10 and 11 show the LDA topics associated with posts published in March and April, respectively.

9 LIWC

In this section, using LIWC (Pennebaker et al., 2015), we identify the LIWC categories associated with posts published in March compared to those published in April in our dataset. Tables 12 and 13 show the LIWC categories associated with posts published in March and April.

10 Discussion and Future Work

This work has several findings; in this section we summarize the results from our analysis.

We find that in posts from our dataset, the emotional support given has a positive correlation with the informational support given; the same applies to the emotional support sought. This implies that when expressing emotional support (giving or seeking), users tend to give informational support as well. Knowing this is important in that it can help the forum moderators and members of the forum

Figure 3: Graph showing the change in the average weekly informational support sought and informational support given in posts from our /r/Coronavirus dataset

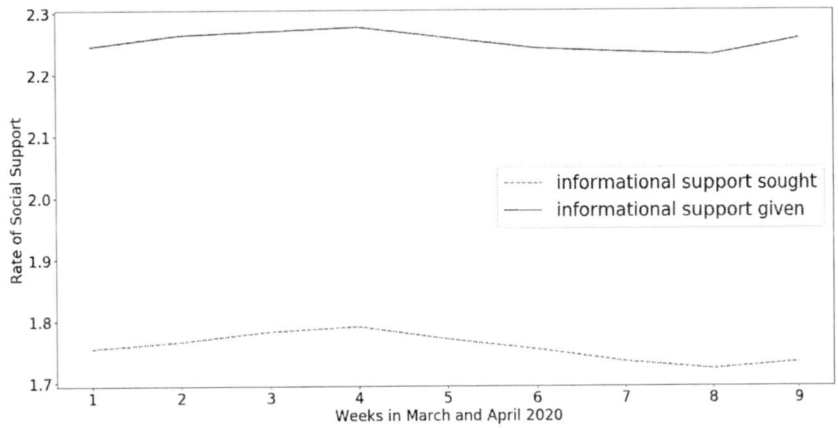

Table 11: Topics associated with posts from our dataset published in April

LDA Topics and highly correlated words	Cohen's D
cases, deaths, rates, number, infected, population, million, total, mortality, Italy	0.181
news, post, find, information, coronavirus, article, link, comments, read, source	0.153
months, normal, lockdown, point, start, vaccine, end, curve, quarantine, life	0.143

Table 12: Results from LIWC categories associated with posts from our dataset published in March

LIWC category	Cohen's D
First person singular pronoun	0.145
Health	0.133
Biological process	0.116
Anxiety	0.115

Table 13: Results from LIWC categories associated with posts from our dataset published in April

LIWC category	Cohen's D
Death	0.155
Anger	0.100

responding to posts to better understand the support needs expressed in these posts, thereby helping inform them on how to best respond to these posts.

Given the scores for emotional support given, emotional support sought, informational support given, and informational support sought for each post published daily in our dataset, using t-test, we determine that the features representing the emotional support given and sought, respectively are statistically different; we do the same for features representing the informational support given and sought, respectively and also determine that these features are statistically different. After determining the statistical difference between these social supports, we then conducted an analysis to determine how the social support expressed in posts in our dataset changes over time i.e. between March and April 2020. We observed that: (i) over time, users sought more emotional support and gave less emotional support and (ii) users gave more informational support compared to informational support sought. These changes in social support needs can provide insights as to the well-being and needs of the public, thereby informing the forum moderators/members and the necessary authorities on the kinds of interventions to provide. For example, given that overtime, users in the forum sought more emotional support, online psychiatry services may be provided. Another finding from this work is that users in the forum tended to give more informational support. We observed that a significant number of posts were relaying information reported by or obtained from other sources such as news articles or websites. In some cases, the veracity of the information given are not determined, hence the potential for the spread of misinformation about the pandemic; this highlights the importance of having experts engage with these online forums in order to provide and verify information.

Given the changes in the expression of social supports (emotional and informational) in our dataset, we identified time periods in which there was a clear distinction in the types of social support expressed in posts. We observed that emotional support sought in posts increased by the beginning of April while the emotional support given decreased. In order to better understand why there was a change in the emotional support sought and given between March and April, using LDA and LIWC, we compared the posts published in this time periods. The intuition for this comparison being that the language use differences in posts published in these time periods may provide insights as to why there was a change in these expressed social supports. We observed that in March, people posted about topics related to traveling and COVID-19 symptoms as shown in Table 10 and used more words associated with the LIWC categories on health and anxiety (Table 12). Also, we observed that in April, people tended to post more about topics related to the numbers of deaths and cases and the lockdown, vaccine, and quarantine (Table 11) and used more words associated with the LIWC categories on death and anger (Table 13). From this language analysis, a potential reason for why the emotional support sought increased in April and the emotional support given decreased is that with the increase in the number of cases and deaths, the lockdown in various places, and the uncertainty of when the pandemic will end and things will go back to normal, people were worried and concerned, hence they sought more emotional support and gave less emotional support. We selected 100 posts with the highest emotional support sought in March and April, respectively, and one of the co-authors went through these posts and it was found that these post were reflective of the LDA and LIWC features that delineate posts published in March and April.

In this work, we tracked social support changes in a 2-months time period; in the future, we aim to track these changes over a longer time period and also track these changes in other Reddit forums focused on discussions around COVID-19. Also, in the future, we aim to analyze COVID-19 related posts in other online platforms such as Twitter, to study the social supports expressed on these platforms and how they vary or are similar from one platform to the other.

Limitations: In this work, we analyzed posts published in the Reddit forum */r/Coronavirus*. Prior work determined that different forums, despite their similarity in topics discussed, attract users with different interests (Tran and Ostendorf, 2016), therefore, the support sought and language used in other COVID-19 related online forums may differ from that of */r/Coronavirus*. Hence the results from our analysis may vary in other COVID-19 online forums or other online platforms.

Ethics and Privacy For all the analysis in this work, we did not use Reddit user names and and no Reddit user was contacted. We had regular weekly meetings with the annotators (who are co-authors of this paper) to address any concerns they might have and to ensure that the content of the data they were labeling did not negatively impact them and their well-being.

11 Conclusion

In this work, we build four machine learning models to measure the extent of the emotional support given, emotional support sought, informational support given, and informational support sought, in posts published on an online forum focused on discussions around COVID-19. We carried out analyses to: (i) determine the correlations between these social supports expressed in posts, (2) determine how these social supports change over time, and (3) using language features, we conducted an analysis to gain insights as to why there was a change in the social supports expressed. The findings from this work show that data from social media can help gauge how people are feeling and responding to the COVID-19 pandemic and track and understand the social support needs of people as it relates to the pandemic. Potentially, these findings can help public health agencies and other agencies design and implement online interventions around COVID-19.

References

Jai Aggarwal, Ella Rabinovich, and Suzanne Stevenson. 2020. Exploration of gender differences in covid-19 discourse on reddit. *arXiv preprint arXiv:2008.05713*.

John W Ayers, Eric C Leas, Derek C Johnson, Adam Poliak, Benjamin M Althouse, Mark Dredze, and Alicia L Nobles. 2020. Internet searches for acute anxiety during the early stages of the covid-19 pandemic. *JAMA Internal Medicine*, 180(12):1706–1707.

John J Bartko. 1966. The intraclass correlation coefficient as a measure of reliability. *Psychological reports*, 19(1):3–11.

David M Blei, Andrew Y Ng, and Michael I Jordan. 2003. Latent dirichlet allocation. *Journal of machine Learning research*, 3(Jan):993–1022.

Emily Chen, Kristina Lerman, and Emilio Ferrara. 2020. Tracking social media discourse about the covid-19 pandemic: Development of a public coronavirus twitter data set. *JMIR Public Health and Surveillance*, 6(2):e19273.

Andre Esteva, Anuprit Kale, Romain Paulus, Kazuma Hashimoto, Wenpeng Yin, Dragomir Radev, and Richard Socher. 2020. Co-search: Covid-19 information retrieval with semantic search, question answering, and abstractive summarization. *arXiv preprint arXiv:2006.09595*.

Sharath Chandra Guntuku, Garrick Sherman, Daniel C Stokes, Anish K Agarwal, Emily Seltzer, Raina M Merchant, and Lyle H Ungar. 2020. Tracking mental health and symptom mentions on twitter during covid-19. *Journal of general internal medicine*, pages 1–3.

Tom Hope, Jason Portenoy, Kishore Vasan, Jonathan Borchardt, Eric Horvitz, Daniel S Weld, Marti A Hearst, and Jevin West. 2020. Scisight: Combining faceted navigation and research group detection for covid-19 exploratory scientific search. *arXiv preprint arXiv:2005.12668*.

Andrew Kachites McCallum. 2002. Mallet: A machine learning for language toolkit. *http://mallet. cs. umass. edu*.

Fabian Pedregosa, Gaël Varoquaux, Alexandre Gramfort, Vincent Michel, Bertrand Thirion, Olivier Grisel, Mathieu Blondel, Peter Prettenhofer, Ron Weiss, Vincent Dubourg, et al. 2011. Scikit-learn: Machine learning in python. *the Journal of machine Learning research*, 12:2825–2830.

James W Pennebaker, Ryan L Boyd, Kayla Jordan, and Kate Blackburn. 2015. The development and psychometric properties of liwc2015. Technical report.

Juan Carlos Medina Serrano, Orestis Papakyriakopoulos, and Simon Hegelich. 2020. Nlp-based feature extraction for the detection of covid-19 misinformation videos on youtube. In *Proceedings of the 1st Workshop on NLP for COVID-19 at ACL 2020*.

Daniel C Stokes, Anietie Andy, Sharath Chandra Guntuku, Lyle H Ungar, and Raina M Merchant. 2020. Public priorities and concerns regarding covid-19 in an online discussion forum: Longitudinal topic modeling. *Journal of general internal medicine*, page 1.

Trang Tran and Mari Ostendorf. 2016. Characterizing the language of online communities and its relation to community reception. *arXiv preprint arXiv:1609.04779*.

Karin Verspoor, Simon Šuster, Yulia Otmakhova, Shevon Mendis, Zenan Zhai, Biaoyan Fang, Jey Han Lau, Timothy Baldwin, Antonio Jimeno Yepes, and David Martinez. 2020. Covid-see: Scientific evidence explorer for covid-19 related research. *arXiv preprint arXiv:2008.07880*.

Lucy Lu Wang, Kyle Lo, Yoganand Chandrasekhar, Russell Reas, Jiangjiang Yang, Darrin Eide, Kathryn Funk, Rodney Kinney, Ziyang Liu, William Merrill, et al. 2020. Cord-19: The covid-19 open research dataset. *ArXiv*.

Yi-Chia Wang, Robert Kraut, and John M Levine. 2012. To stay or leave? the relationship of emotional and informational support to commitment in online health support groups. In *Proceedings of the ACM 2012 conference on computer supported cooperative work*, pages 833–842.

Jerry Wei, Chengyu Huang, Soroush Vosoughi, and Jason Wei. 2020. What are people asking about covid-19? a question classification dataset. *arXiv preprint arXiv:2005.12522*.

Theresa Wilson, Janyce Wiebe, and Paul Hoffmann. 2005. Recognizing contextual polarity in phrase-level sentiment analysis. In *Proceedings of human language technology conference and conference on empirical methods in natural language processing*, pages 347–354.

Diyi Yang, Robert Kraut, and John M Levine. 2017. Commitment of newcomers and old-timers to online health support communities. In *Proceedings of the 2017 CHI conference on human factors in computing systems*, pages 6363–6375.

Jason Shuo Zhang, Brian C Keegan, Qin Lv, and Chenhao Tan. 2020. A tale of two communities: Characterizing reddit response to covid-19 through/r/china_flu and/r/coronavirus. *arXiv preprint arXiv:2006.04816*.

Shi Zong, Ashutosh Baheti, Wei Xu, and Alan Ritter. 2020. Extracting covid-19 events from twitter. *arXiv preprint arXiv:2006.02567*.

Fast and Effective Biomedical Entity Linking Using a Dual Encoder

Rajarshi Bhowmik and **Karl Stratos**
Rutgers University–New Brunswick
Piscataway, New Jersey, USA
rajarshi.bhowmik@rutgers.edu
karl.stratos@rutgers.edu

Gerard de Melo
Hasso Plattner Institute
University of Potsdam
Potsdam, Germany
gdm@demelo.org

Abstract

Biomedical entity linking is the task of identifying mentions of biomedical concepts in text documents and mapping them to canonical entities in a target thesaurus. Recent advancements in entity linking using BERT-based models follow a *retrieve and rerank* paradigm, where the candidate entities are first selected using a retriever model, and then the retrieved candidates are ranked by a reranker model. While this paradigm produces state-of-the-art results, they are slow both at training and test time as they can process only one mention at a time. To mitigate these issues, we propose a BERT-based dual encoder model that resolves multiple mentions in a document in one shot. We show that our proposed model is multiple times faster than existing BERT-based models while being competitive in accuracy for biomedical entity linking. Additionally, we modify our dual encoder model for end-to-end biomedical entity linking that performs both mention span detection and entity disambiguation and out-performs two recently proposed models.

1 Introduction

Entity linking is the task of identifying mentions of named entities (or other terms) in a text document and disambiguating them by mapping them to canonical entities (or concepts) listed in a reference knowledge graph (Hogan et al., 2020). This is an essential step in information extraction, and therefore has been studied extensively both in domain-specific and domain-agnostic settings. Recent state-of-the-art models (Logeswaran et al., 2019; Wu et al., 2019) attempt to learn better representations of mentions and candidates using the rich contextual information encoded in pre-trained language models such as BERT (Devlin et al., 2019). These models follow a *retrieve and rerank* paradigm, which consists of two separate steps: First, the candidate entities are selected using a retrieval model. Subsequently, the retrieved candidates are ranked by a reranker model.

Although this approach has yielded strong results, owing primarily to the powerful contextual representation learning ability of BERT-based encoders, these models typically process a single mention at a time. Processing one mention at a time incurs a substantial overhead both during training and test time, leading to a system that is slow and impractical.

In this paper, we propose a *collective entity linking* method that processes an entire document only once, such that all entity mentions within it are linked to their respective target entities in the knowledge base in one pass.

Compared to the popular entity linking model BLINK (Wu et al., 2019), our model is up to 25x faster. BLINK deploys two separately trainable models for candidate retrieval and reranking. In contrast, our method learns a single model that can perform both the retrieval and reranking steps of entity linking. Our model does not require candidate retrieval at inference time, as our dual encoder approach allows us to compare each mention to all entities in the target knowledge base, thus significantly reducing the overhead at inference time.

We evaluate our method on two particularly challenging datasets from the biomedical domain. In recent times, there is an increased focus on information extraction from biomedical text such as biomedical academic publications, electronic health records, discharge summaries of patients, or clinical reports. Extracting named concepts from biomedical text requires domain expertise. Existing automatic extraction methods, including the methods and tools catering to the biomedical domain (Savova et al., 2010; Soldaini and Goharian, 2016; Aronson, 2006), often perform poorly due to the inherent challenges of biomedical text:

Proceedings of the 12th International Workshop on Health Text Mining and Information Analysis, pages 28–37
April 19, 2021.

(1) Biomedical text typically contains substantial domain-specific jargon and abbreviations. For example, *CT* could stand for *Computed tomography* or *Copper Toxicosis*. (2) The target concepts in the knowledge base often have very similar surface forms, making the disambiguation task difficult. For example, *Pseudomonas aeruginosa* is a kind of bacteria, while *Pseudomonas aeruginosa infection* is a disease. Many existing biomedical information extraction tools rely on similarities in surface forms of mentions and candidates, and thus invariably falter in more challenging cases such as these. Additionally, long mention spans (e.g., disease names) and the density of mentions per document make the biomedical entity linking very challenging.

Contributions The key contributions of our work are as follows.

- Training our collective entity disambiguation model is 3x faster than other dual encoder models with the same number of parameters that perform per-mention entity disambiguation. At inference time, our model is 3-25x faster than other comparable models.
- At the same time, our model obtains favorable results on two biomedical datasets compared to state-of-the-art entity linking models.
- Our model can also perform end-to-end entity linking when trained with the multi-task objective of mention span detection and entity disambiguation. We show that without using any semantic type information, our model significantly out-performs two recent biomedical entity linking models – MedType (Vashishth et al., 2020) and SciSpacy (Neumann et al., 2019) – on two benchmark datasets.

2 Related Work

2.1 Entity Linking

The task of entity linking has been studied extensively in the literature. In the past, most models relied on hand-crafted features for entity disambiguation using surface forms and alias tables, which may not be available for every domain. With the advent of deep learning, contextual representation learning for mention spans has become more popular. Recent Transformer-based models for entity linking (Wu et al., 2019; Févry et al., 2020) have achieved state-of-the-art performance on traditional benchmark datasets such as AIDA-CoNLL and TACKBP 2010.

2.2 Biomedical Entity Linking

In the biomedical domain, there are many existing tools, such as TaggerOne (Leaman and Lu, 2016), MetaMap (Aronson, 2006), cTAKES (Savova et al., 2010), QuickUMLS (Soldaini and Goharian, 2016), among others, for normalizing mentions of biomedical concepts to a biomedical thesaurus. Most of these methods rely on feature-based approaches. Recently, Zhu et al. (2019) proposed a model that utilizes the latent semantic information of mentions and entities to perform entity linking. Other recent models such as Xu et al. (2020) and Vashishth et al. (2020) also leverage semantic type information for improved entity disambiguation. Our work is different from these approaches, as our model does not use semantic type information, since such information may not always be available. Recent studies such as Xu et al. (2020) and Ji et al. (2020) deploy a BERT-based retrieve and re-rank model. In contrast, our model does not rely on a separate re-ranker model, which significantly improves its efficiency.

2.3 End-to-End Entity Linking

End-to-end entity linking refers to the task of predicting mention spans and the corresponding target entities jointly using a single model. Traditionally, span detection and entity disambiguation tasks were done in a pipelined approach, making these approaches susceptible to error propagation. To alleviate this issue, Kolitsas et al. (2018) proposed a neural end-to-end model that performs the dual tasks of mention span detection and entity disambiguation. However, for span detection and disambiguation, their method relies on an empirical probabilistic entity mapping $p(e|m)$ to select a candidate set $C(m)$ for each mention m. Such mention–entity prior $p(e|m)$ is not available in every domain, especially in the biomedical domain that we consider in this paper. In contrast, our method does not rely on any extrinsic sources of information. Recently, Furrer et al. (2020) proposed a parallel sequence tagging model that treats both span detection and entity disambiguation as sequence tagging tasks. However, one practical disadvantage of their model is the large number of tag labels when the target knowledge base contains thousands of entities. In contrast, our dual encoder model can effectively link mentions to a knowledge base with large number of entities.

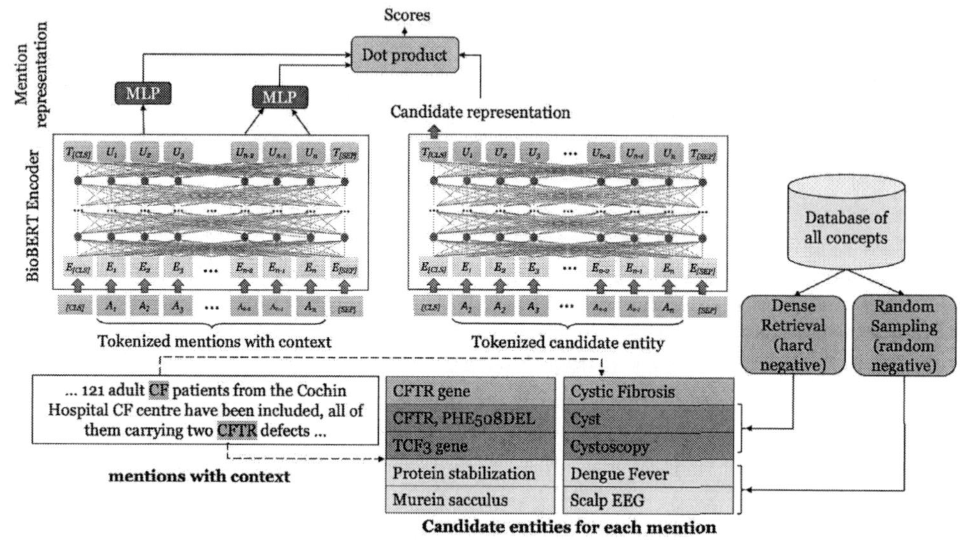

Figure 1: A schematic diagram of the Dual Encoder model for collective entity disambiguation. In this diagram, the number of mentions in a document and the number of candidate entities per mention are for illustration purpose only. The inputs to the BioBERT encoders are the tokens obtained from the BioBERT tokenizer.

3 Model

Given a document $d = [x_1^d, \ldots, x_T^d]$ of T tokens with N mentions $\{m_1, \ldots, m_N\}$ and a set of M entities $\{e_1, \ldots, e_M\}$ in a target knowledge base or thesaurus $\mathcal{E}$, the task of collective entity disambiguation consists in mapping each entity mention m_k in the document to a target entity $t_k \in \mathcal{E}$ in one shot. Each mention in the document d may span over one or multiple tokens, denoted by pairs (i, j) of start and end index positions such that $m_k = [x_i^d, \ldots, x_j^d]$.

3.1 Encoding Mentions and Candidates

Our model consists of two BERT-based encoders. The *mention encoder* is responsible for learning representations of contextual mentions and the *candidate encoder* learns representations for the candidate entities. A schematic diagram of the model is presented in Figure 1. Following the BERT model, the input sequences to these encoders start and end with the special tokens `[CLS]` and `[SEP]`, respectively.

Mention Encoder Given an input text document $[x_1^d, \ldots, x_T^d]$ of T tokens with M mentions, the output of the final layer of the encoder, denoted by $[\mathbf{h_1}, \ldots, \mathbf{h_T}]$, is a contextualized representation of the input tokens. For each mention span (i, j), we concatenate the first and the last tokens of the span and pass it through a linear layer to obtain the representations for each of the mentions. Formally, the representation of mention m_k is given as

$$\mathbf{u_k^m} = \mathbf{W}[\mathbf{h_i}; \mathbf{h_j}] + \mathbf{b}. \tag{1}$$

Since the encoder module deploys a self-attention mechanism, every mention inherently captures contextual information from the other mentions in the document.

Candidate Encoder Given an input candidate entity $e = [y_1^e, \ldots, y_T^e]$ of T tokens, the output of the final layer corresponding to the `[CLS]` token yields the representation for the candidate entity. We denote the representation of entity e as $\mathbf{v^e}$. As shown in Figure 1, we use the UMLS concept name of each candidate entity as the input to the candidate encoder.

3.2 Candidate Selection

Candidate Retrieval Since the entity disambiguation task is formulated as a learning to rank problem, we need to retrieve negative candidate entities for ranking during training. To this end, we randomly sample a set of negative candidates from the pool of all entities in the knowledge base. Additionally, we adopt the *hard negative mining* strategy used by Gillick et al. (2019) to retrieve negative candidates by performing nearest neighbor search using the dense representations of mentions and candidates described above. The hard negative candidates are the entities that are more similar to the mention than the gold target entity.

Candidate Scoring The retrieved set of candidate entities $\mathcal{C}_k = \{c_1^k, \ldots, c_l^k\}$ for each mention m_k are scored using a dot product between the mention representation $\mathbf{u_k^m}$ and each candidate representation $\mathbf{v^c}$. Formally, for each $c \in \mathcal{C}_k$

$$\psi(m_k, c) = (\mathbf{u_k^m})^\intercal \mathbf{v^c} \tag{2}$$

3.3 Training and Inference

Loss Function and Training We train our model using the cross-entropy loss function to maximize the score of the gold target entities.

Inference During inference, we do not require candidate retrieval per mention. The representations of all entities in the knowledge base $\mathcal{E}$ can be pre-computed and cached. The inference task is thus reduced to finding the maximum dot product between each mention representation and all entity representations.

$$\hat{t}_k = \arg\max_{e \in \mathcal{E}} \{(\mathbf{u_k^m})^\intercal \mathbf{v^e}\} \tag{3}$$

3.4 End-to-End Entity Linking

Many of the state-of-the-art entity disambiguation models assume that gold mention spans are available during test time and thus have limited applicability in real-world entity linking tasks, where such gold mentions are typically not available. To avoid this, recent works (Kolitsas et al., 2018; Févry et al., 2020; Li et al., 2020) have investigated end-to-end entity linking, where a model needs to perform both mention span detection and entity disambiguation.

Mention Span Detection We experiment with two different methods for mention span detection with different computational complexity. In our first method, following Févry et al. (2020), we use a simple `BIO` tagging scheme to identify the mention spans. Every token in the input text is annotated with one of these three tags. Under this tagging scheme, any contiguous segment of tokens starting with a `B` tag and followed by `I` tags is treated as a mention. Although this method is computationally efficient ($\mathcal{O}(T)$), our empirical results suggest that it is not as effective as the following.

Following the recent work of Kolitsas et al. (2018) and Li et al. (2020), our mention span detection method enumerates all possible spans in the input text document as potential mentions. However, enumerating all possible spans in a document of length T is prohibitively large ($\mathcal{O}(T^2)$) and computationally expensive. Therefore, we constrain the maximum length of a mention span to $L \ll T$.

We calculate the probability of each candidate mention span (i, j) as follows.

$$p(m|(i,j)) = \sigma(\mathbf{w_s^\intercal h_i} + \mathbf{w_e^\intercal h_j} + \sum_{q=i}^{j} \mathbf{w_m^\intercal h_q}) \tag{4}$$

where $\mathbf{w_s}$, $\mathbf{w_e}$, and $\mathbf{w_m}$ are trainable parameters and $\sigma(x) = \frac{1}{1+e^{-x}}$.

Entity Disambiguation We represent each mention (i, j) by mean pooling the final layer of the encoder, i.e., $\mathbf{u^m_{(i,j)}} = \frac{1}{j-i+1} \sum_{q=i}^{j} \mathbf{h_q}$. During training, we perform candidate selection as described in Section 3.2.

We jointly train the model by minimizing the sum of mention detection loss and entity disambiguation loss. We use a binary cross-entropy loss for mention detection with the gold mention spans as positive and other candidate mention spans as negative samples. For entity disambiguation, we use the cross-entropy loss to minimize the negative log likelihood of the gold target entity given a gold mention span.

During inference, we choose only the candidate mentions with $p(m|(i,j)) > \gamma$ as the predicted mention spans. Then, as mentioned in Section 3.3, we determine the maximum dot product between the mention representations and all candidate entity representations to predict the entity for each predicted mention during inference.

4 Evaluation

4.1 Datasets

Our experiments are conducted on two challenging datasets from the biomedical domain – MedMentions (Mohan and Li, 2019) and the BioCreative V Chemical Disease Relation (BC5CDR) dataset (Li et al., 2016). In the following, we provide some details of these two datasets, while basic statistics are given in Table 1.

MedMentions is a large-scale biomedical corpus annotated with UMLS concepts. It consists of a total of $4,392$ English language abstracts published on PubMed®. The dataset has $352,496$ mentions, and each mention is associated with a single UMLS Concept Unique Identifier (CUI) and one or more

Datasets	Mentions	Mentions/Doc	Unique Concepts	Types
MedMentions	352,496	80	34,724	128
BC5CDR	28,559	19	9,149	2

Table 1: Details of the datasets used for evaluation.

semantic types identified by a Type Unique Identifier (TUI). The concepts belong to 128 different semantic types. MedMentions also provides a 60% – 20% – 20% random partitioning of the corpus into training, development, and test sets. Note that 12% of the concepts in the test dataset do not occur in the training or development sets. For this dataset, our target KB consists of the concepts that are linked to at least one mention in the MedMentions dataset.

The BC5CDR corpus consists of 1,500 English language PubMed® articles with 4,409 annotated chemicals and 5,818 diseases, which are equally partitioned into training, development, and test sets. Each entity annotation includes both the mention text spans and normalized concept identifiers, using MeSH as the target vocabulary. Apart from entity linking annotations, this dataset also provides 3,116 chemical–disease relations. However, identifying relations between mentions is beyond the scope of our study on entity linking and hence, we ignore these annotations.

4.2 Baselines

We compare our model against some of the recent state-of-the-art entity linking models from both the biomedical and non-biomedical domains. In the biomedical domain, LATTE (Zhu et al., 2019) showed state-of-the-art results on the MedMentions dataset. However, we find that LATTE adds the gold target entity to the set of candidates retrieved by the BM25 retrieval method during both training and inference.

The Cross Encoder model proposed by Logeswaran et al. (2019), which follows a *retrieve and rerank* paradigm, has been successfully adopted in the biomedical domain by Xu et al. (2020) and Ji et al. (2020). This model uses a single encoder. The input to this encoder is a concatenation of a mention with context and a candidate entity with a `[SEP]` token in between. This allows cross-attention between mentions and candidate entities. We use our own implementation of the model by Logeswaran et al. (2019) for comparison.

We also compare with BLINK (Wu et al., 2019), a state-of-the-art entity linking model that uses dense retrieval using dual encoders for candidate generation, followed by a cross-encoder for reranking.

Additionally, we use the dual encoder model that processes each mention independently as a baseline. In principle, this baseline is similar to the retriever model of Wu et al. (2019) and Gillick et al. (2019).

For the task of end-to-end entity disambiguation, we compare our models with two recent state-of-the-art models – SciSpacy (Neumann et al., 2019) and MedType (Vashishth et al., 2020). SciSpacy uses overlapping character N-grams for mention span detection and entity disambiguation. MedType improves the results of SciSpacy by using a better candidate retrieval method that exploits the semantic type information of the candidate entities.

4.3 Experimental Details

In this section, we provide details pertaining to the experiments for the purpose of reproducibility. We also make the code publicly available [1].

Domain-Adaptive Pretraining Recent studies (Logeswaran et al., 2019; Févry et al., 2020; Wu et al., 2019) have shown that pre-training BERT on the target domain provides additional performance gains for entity linking. Following this finding, we adopt BioBERT (Lee et al., 2019) as our domain-specific pretrained model. BioBERT is intitialzed with the parameters of the original BERT model, and further pretrained on PubMed abstracts to adapt to biomedical NLP tasks.

Data Wrangling In theory, our collective entity disambiguation model is capable of processing documents of arbitrary length. However, there are practical constraints. First, the GPU memory limit enforces an upper bound on the number of mentions that can be processed together, and secondly, BERT stipulates the maximum length of the input sequence to be 512 tokens. To circumvent these constraints, we segment each document so that each chunk contains a maximum of 8 mentions or a maximum of 512 tokens (whichever happens earlier). After this data wrangling process, the 4,392

[1] https://github.com/kingsaint/BioMedical-EL

Model	Candidate retrieval method		Unnormalized		Normalized	
	Training	Test	P@1	MAP	P@1	MAP
† LATTE	BM25	BM25	-	-	88.5	92.8
† Cross Encoder	BM25	BM25	-	-	91.6	95.1
Cross Encoder	BM25	BM25	53.8	56.2	90.4	94.4
Dual Encoder (1 mention)	DR (random)	all entities	54.1	64.8	N/A	N/A
Dual Encoder (1 mention)	DR (random + hard)	all entities	62.9	69.7	N/A	N/A
BLINK	DR (random + hard)	DR (hard)	68.1	73.0	84.7	90.8
Dual Encoder (collective)	DR (random)	all entities	58.2	68.5	N/A	N/A
Dual Encoder (collective)	DR (random + hard)	all entities	**68.4**	**75.6**	N/A	N/A

Table 2: Precision@1 and Mean Average Precision (MAP) for the entity disambiguation task on the MedMentions dataset when the gold mention spans are known. † LATTE results are copied from the original paper and always incorporate gold entities as candidates (thus recall is always 100%). † Cross Encoder shows results in this setting as a reference point. Models without † do not add gold entities to the candidate set. 'N/A' stands for 'Not Applicable'. 'DR' stands for dense retrieval.

Model	Candidate retrieval method		Unnormalized		Normalized	
	Training	Test	P@1	MAP	P@1	MAP
Cross Encoder	BM25	BM25	72.1	73.1	96.8	98.1
Dual Encoder (1 mention)	DR (random)	all entities	76.3	82.4	N/A	N/A
Dual Encoder (1 mention)	DR (random + hard)	all entities	**84.8**	**87.7**	N/A	N/A
BLINK	DR (random + hard)	DR (hard)	74.7	75.6	97.2	98.4
Dual Encoder (collective)	DR (random)	all entities	69.0	77.2	N/A	N/A
Dual Encoder (collective)	DR (random + hard)	all entities	80.7	85.1	N/A	N/A

Table 3: Precision@1 and Mean Average Precision (MAP) for the entity disambiguation task on the BC5CDR dataset when the gold mention spans are known. 'N/A' stands for 'Not Applicable'. 'DR' stands for dense retrieval.

original documents in the MedMentions dataset are split into $44,983$ segmented documents. Note that during inference our model can process more than 8 mentions. However, without loss of generality, we assumed the same segmentation method during inference. We postulate that with more GPU memory and longer context (Beltagy et al., 2020), our collective entity disambiguation model will be able to process documents of arbitrary length without segmentation during training and inference.

For the other baselines, we process each mention along with its contexts independently. We found that a context window of 128 characters surrounding each mention suffices for these models. We also experimented with longer contexts and observed that the performance of the models deteriorates.

Hyperparameters To encode mentions, we use a context window of up to 128 tokens for the single-mention Dual Encoder. The candidate entities are tokenized to a maximal length of 128 tokens across all Dual Encoder models. In the Cross Encoder and BLINK models, where candidate tokens are appended to the context tokens, we use a maximum of 256 tokens. For Collective Dual Encoder models, the mention encoder can encode a tokenized document of maximum length 512. For all our experiments, we use AdamW stochastic optimization and a linear scheduling for the learning rate of the optimizer. For the single-mention Dual Encoder, Cross Encoder and BLINK model, we find an initial learning rate of 0.00005 to be optimal. For collective Dual Encoder models, we find an initial learning rate of 0.00001 to be suitable for both the end-to-end and non-end-to-end settings. The ratio of hard and random negative candidates is set to 1:1, as we choose 10 samples from each. For each model, the hyperparameters are tuned using the validation set. For the end-to-end entity linking model, we set the maximum length of a mention span L to 10 tokens.

4.4 Evaluation Metrics

Picking the correct target entity among a set of candidate entities is a learning to rank problem. Therefore, we use Precision@1 and Mean Average Precision (MAP) as our evaluation metrics when the gold mention spans are known. Since there is only one correct target entity per mention in our datasets, Precision@1 is also equivalent to the accuracy. One can consider these metrics in normalized and unnormalized settings. The normalized setting is applicable when candidate retrieval is done during inference and the target entity is present in the set of retrieved candidates. Since our model and other Dual Encoder based models do not require retrieval at test time, the normalized evaluation setting is not applicable in these cases.

4.5 Results

Entity Disambiguation We provide the results of our experiments for the entity disambiguation task on the MedMentions and BC5CDR datasets in Tables 2 and 3, respectively. For the MedMentions dataset, our collective dual encoder model outperforms all other models, while being extremely time efficient during training and inference. On the BC5CDR dataset, our method performs adequately as compared to other baselines. Our model compares favorably against the state-of-the-art entity linking model BLINK on both datasets. Surprisingly, for the BC5CDR dataset, BLINK is outperformed by the Dual Encoder baselines that process each mention independently, despite the fact that BLINK's input candidates are generated by this model. We conjecture that BLINK's cross encoder model for re-ranking is more susceptible to overfitting on this relatively small-scale dataset. Our model consistently outperforms the Cross Encoder model, which reinforces the prior observations made by Wu et al. (2019) that dense retrieval of candidates improves the accuracy of entity disambiguation models. Finally, comparisons with an ablated version of our model that uses only random negative candidates during training show that hard negative mining is essential for the model for better entity disambiguation.

Training and Inference Speed We perform a comparative analysis of the training speed of our collective dual encoder model with the single-mention dual encoder model. We show in Fig. 2 and 3 that our model achieves higher accuracy and recall@10 much faster than the single-mention dual encoder model. In fact, our model is 3x faster than the single-mention Dual Encoder model.

We also compare the inference speed of our model with BLINK and the single-mention Dual Encoder model. The comparisons of inference

Model	mentions/sec
BLINK	11.5
Dual Encoder (1 mention)	65.0
Dual Encoder (collective)	**192.4**

Table 4: Inference speed comparison on the MedMentions dataset.

Model	mentions/sec
BLINK	11.5
Dual Encoder (1 mention)	87.0
Dual Encoder (collective)	**402.5**

Table 5: Inference speed comparison on the BC5CDR dataset.

speed for the two datasets are presented in Tables 4 and 5, respectively. The inference speed is measured on a single NVIDIA Quadro RTX GPU with batch size 1. We observe that our collective dual encoder model is 3-4x faster than the single-mention Dual Encoder model and up to 25x faster (on average over the two datasets) than BLINK. Since our model can process a document with N mentions in one shot, we achieve higher entity disambiguation speed than the single-mention Dual Encoder and the BLINK model – both require N forward passes to process the N mentions in a document. For these experiments, we set $N = 8$, i.e., our collective dual encoder model processes up to 8 mentions in a single pass. Note that the value of N could be increased further for the inference phase. Caching the entity representations also helps our model and the single-mention Dual Encoder model at test time. The cross encoder of BLINK prevents it from using any cached entity representations, which drastically slows down the entity resolution speed of BLINK.

Candidate Recall We compare the recall@10 metrics of BM25 retrieval method used in LATTE and Cross Encoder to the dense retrieval method used in BLINK and in our model. We present our results in Tables 6 for the MedMentions and BC5CDR datasets, respectively. Similar to the observations made for BLINK and Gillick et al. (2019), we also find that dense retrieval has a superior recall than BM25. However, we observe that the recall value of dense retrieval depends on the underlying entity disambiguation model. For instance, on the MedMentions dataset, our model has much higher recall@10 than the Dual Encoder model that processes each mention independently,

	MedMentions		BC5CDR	
Model	Dev	Test	Dev	Test
BM25	59.8	59.5	76.3	74.5
Dense retrieval (1 mention)	80.2	80.6	92.1	**92.3**
Dense retrieval (collective)	**87.5**	**87.6**	**92.2**	**92.3**

Table 6: Comparison of development and test set Recall@10 on MedMentions and BC5CDR datasets

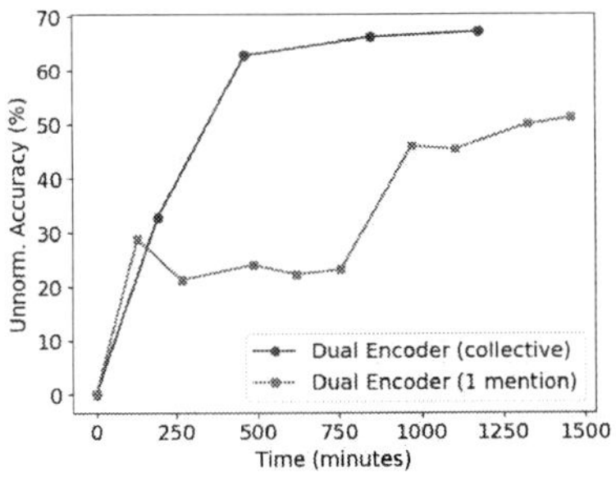

Figure 2: Comparative analysis of training speed measured in terms of accuracy achieved in first 24 hours of training. Both models were trained on 4 NVIDIA Quadro RTX GPUs with 24 GB memory.

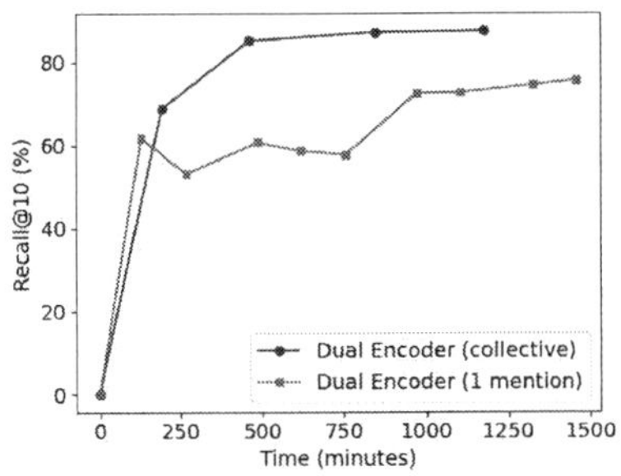

Figure 3: Comparative analysis of training speed measured in terms of recall@10 achieved in first 24 hours of training. Both models were trained on 4 NVIDIA Quadro RTX GPUs with 24 GB memory.

while both models are trained using a combination of hard and random negative candidates. However, this observation is not consistent across datasets as we do not observe similar gains in recall@10 for the BC5CDR dataset. We will explore this phenomenon in future work.

End-to-End Entity Disambiguation For the end-to-end entity linking task, we evaluate the models with two different evaluation protocols. In the *strict match* protocol, the predicted mention spans and predicted target entity must match strictly with the gold spans and target entity. In the *partial match* protocol, if there is an overlap between the predicted mention span and the gold mention span, and the predicted target entity matches the gold target entity, then it is considered to be a true positive. We evaluate our models using micro-averaged precision, recall, and F1 scores as evaluation metrics. For a fair comparison, we use the off-the-shelf evaluation tool `neleval`[2], which is also used for MedType. We follow the same evaluation protocol and settings as used for MedType.

We present the results of our collective Dual Encoder model and the baselines in Table 7. The results show that exhaustive search over all possible spans for mention detection yields significantly better results than the `BIO` tagging based method, despite the additional computational cost. Moreover, our dual encoder based end-to-end entity linking model significantly outperforms SciSpacy and MedType. Note that there are highly specialized models such as TaggerOne (Leaman and Lu, 2016) that perform much better than our model on the BC5CDR dataset. However, TaggerOne is suitable for a few specific types of entities such as *Disease* and *Chemical*. For a dataset with entities of various different semantic types (e.g., MedMentions), Mohan and Li (2019) show that TaggerOne performs inadequately. For such datasets where the target entities belong to many different semantic types, our proposed model is more effective as compared to highly specialized models like TaggerOne.

5 Conclusion

This paper introduces a biomedical entity linking approach using BERT-based dual encoders to disambiguate multiple mentions of biomedical concepts in a document in a single shot. We show empirically that our method achieves higher accuracy and recall than other competitive baseline models in significantly less training and inference time. We also showed that our method is significantly better than two recently proposed biomedical entity linking models for the end-to-end entity disambiguation task when subjected to multi-task

[2] https://github.com/wikilinks/neleval

Model	MedMentions						BC5CDR					
	Partial match			Strict match			Partial match			Strict match		
	P	R	F1	P	R	F1	P	R	F1	P	R	F1
SciSpacy	40.9	40.2	40.6	37.7	36.6	37.1	15.5	53.4	24.0	14.5	48.4	22.3
MedType	44.7	44.1	44.4	41.2	40.0	40.6	16.6	57.0	25.7	15.3	51.0	23.5
Dual Encoder (BIO tags)	44.5	37.6	40.7	41.2	34.9	37.8	29.2	31.5	30.3	10.2	10.8	10.5
Dual Encoder (Exhaustive)	**56.3**	**56.4**	**56.4**	**52.9**	**53.8**	**53.4**	**76.0**	**74.4**	**75.2**	**74.6**	**73.1**	**73.8**

Table 7: Micro Precision (P), Recall (R) and F1 scores for the end-to-end entity linking task on the MedMentions and BC5DCR datasets.

learning objectives for joint mention span detection and entity disambiguation using a single model.

Acknowledgments

We thank Vipina Kuttichi Keloth for her generous assistance in data processing and initial experiments. We thank Diffbot and the Google Cloud Platform for granting us access to computing infrastructure used to run some of the experiments reported in this paper.

References

Alan R Aronson. 2006. Metamap: Mapping text to the umls metathesaurus. *Bethesda, MD: NLM, NIH, DHHS*, 1:26.

Iz Beltagy, Matthew E. Peters, and Arman Cohan. 2020. Longformer: The long-document transformer. *arXiv:2004.05150*.

Jacob Devlin, Ming-Wei Chang, Kenton Lee, and Kristina Toutanova. 2019. BERT: Pre-training of deep bidirectional transformers for language understanding. In *Proceedings of the 2019 Conference of the North American Chapter of the Association for Computational Linguistics: Human Language Technologies, Volume 1 (Long and Short Papers)*, pages 4171–4186, Minneapolis, Minnesota. Association for Computational Linguistics.

Lenz Furrer, Joseph Cornelius, and Fabio Rinaldi. 2020. Parallel sequence tagging for concept recognition.

Thibault Févry, Nicholas FitzGerald, Livio Baldini Soares, and Tom Kwiatkowski. 2020. Empirical evaluation of pretraining strategies for supervised entity linking.

Daniel Gillick, Sayali Kulkarni, Larry Lansing, Alessandro Presta, Jason Baldridge, Eugene Ie, and Diego Garcia-Olano. 2019. Learning dense representations for entity retrieval. In *Proceedings of the 23rd Conference on Computational Natural Language Learning (CoNLL)*, pages 528–537, Hong Kong, China. Association for Computational Linguistics.

Aidan Hogan, Eva Blomqvist, Michael Cochez, Claudia d'Amato, Gerard de Melo, Claudio Gutierrez, José Emilio Labra Gayo, Sabrina Kirrane, Sebastian Neumaier, Axel Polleres, Roberto Navigli, Axel-Cyrille Ngonga Ngomo, Sabbir M. Rashid, Anisa Rula, Lukas Schmelzeisen, Juan Sequeda, Steffen Staab, and Antoine Zimmermann. 2020. Knowledge graphs. *ArXiv*, 2003.02320.

Zongcheng Ji, Qiang Wei, and Hua Xu. 2020. Bert-based ranking for biomedical entity normalization. *AMIA Summits on Translational Science Proceedings*, 2020:269.

Nikolaos Kolitsas, Octavian-Eugen Ganea, and Thomas Hofmann. 2018. End-to-end neural entity linking. In *Proceedings of the 22nd Conference on Computational Natural Language Learning*, pages 519–529, Brussels, Belgium. Association for Computational Linguistics.

Robert Leaman and Zhiyong Lu. 2016. Taggerone: joint named entity recognition and normalization with semi-markov models. *Bioinformatics*, 32(18):2839–2846.

Jinhyuk Lee, Wonjin Yoon, Sungdong Kim, Donghyeon Kim, Sunkyu Kim, Chan Ho So, and Jaewoo Kang. 2019. Biobert: a pre-trained biomedical language representation model for biomedical text mining. *Bioinformatics*.

Belinda Z. Li, Sewon Min, Srinivasan Iyer, Yashar Mehdad, and Wen-tau Yih. 2020. Efficient one-pass end-to-end entity linking for questions. In *Proceedings of the 2020 Conference on Empirical Methods in Natural Language Processing (EMNLP)*, pages 6433–6441, Online. Association for Computational Linguistics.

Jiao Li, Yueping Sun, Robin J Johnson, Daniela Sciaky, Chih-Hsuan Wei, Robert Leaman, Allan Peter Davis, Carolyn J Mattingly, Thomas C Wiegers, and Zhiyong Lu. 2016. Biocreative v cdr task corpus: a resource for chemical disease relation extraction. *Database*, 2016.

Lajanugen Logeswaran, Ming-Wei Chang, Kenton Lee, Kristina Toutanova, Jacob Devlin, and Honglak Lee. 2019. Zero-shot entity linking by reading entity descriptions. In *Proceedings of the 57th Annual Meeting of the Association for Computational Linguistics*, pages 3449–3460, Florence, Italy. Association for Computational Linguistics.

Sunil Mohan and Donghui Li. 2019. Medmentions: A large biomedical corpus annotated with umls concepts. *ArXiv*, abs/1902.09476.

Mark Neumann, Daniel King, Iz Beltagy, and Waleed Ammar. 2019. ScispaCy: Fast and robust models for biomedical natural language processing. In *Proceedings of the 18th BioNLP Workshop and Shared Task*, pages 319–327, Florence, Italy. Association for Computational Linguistics.

Guergana K Savova, James J Masanz, Philip V Ogren, Jiaping Zheng, Sunghwan Sohn, Karin C Kipper-Schuler, and Christopher G Chute. 2010. Mayo clinical text analysis and knowledge extraction system (ctakes): architecture, component evaluation and applications. *Journal of the American Medical Informatics Association*, 17(5):507–513.

Luca Soldaini and Nazli Goharian. 2016. Quickumls: a fast, unsupervised approach for medical concept extraction. In *MedIR workshop, sigir*, pages 1–4.

Shikhar Vashishth, Rishabh Joshi, Ritam Dutt, Denis Newman-Griffis, and Carolyn Rose. 2020. Med-Type: Improving Medical Entity Linking with Semantic Type Prediction. *arXiv e-prints*, page arXiv:2005.00460.

Ledell Wu, Fabio Petroni, Martin Josifoski, Sebastian Riedel, and Luke Zettlemoyer. 2019. Zero-shot entity linking with dense entity retrieval. In *arXiv:1911.03814*.

Dongfang Xu, Zeyu Zhang, and Steven Bethard. 2020. A generate-and-rank framework with semantic type regularization for biomedical concept normalization. In *Proceedings of the 58th Annual Meeting of the Association for Computational Linguistics*, pages 8452–8464, Online. Association for Computational Linguistics.

Ming Zhu, Busra Celikkaya, Parminder Bhatia, and Chandan K. Reddy. 2019. Latte: Latent type modeling for biomedical entity linking.

Leveraging knowledge sources for detecting self-reports of particular health issues on social media

Parsa Bagherzadeh and Sabine Bergler
CLaC Labs, Concordia University
Montreal, Canada
{p_bagher, bergler}@cse.concordia.ca

Abstract

This paper investigates incorporating quality knowledge sources developed by experts for the medical domain as well as syntactic information for classification of tweets into four different health oriented categories. We claim that resources such as the MeSH hierarchy and currently available parse information are effective extensions of moderately sized training datasets for various fine-grained tweet classification tasks of self-reported health issues.

1 Introduction

Social media are a ubiquously accessible way to communicate and interact with others, making their users producers of Big Data at a fast rate. It is estimated that about 500M tweets are sent each day on Twitter which often contain information about opinions, trends, reviews, health, incidents, etc. This offers the possibility to gain insight into individuals' behavior and general state in direct and unmitigated fashion (Rousidis et al., 2020).

Health applications based on social media are an active research area for outbreak management, disease surveillance (Charles-Smith et al., 2015), and pharmacovigilance (Golder et al., 2015). For instance, epidemiologists hope to mine social media to predict and monitor the likelihood and possible severity of outbreaks in a timely fashion. Systems that support this type of research have to make predictions from incomplete data of varying quality.

Deep learning methods are popular for NLP applications and demonstrated significant improvements in areas such as text classification. Deep models have been widely used for personal health mention detection (Khan et al., 2020), (Sarabadani, 2019), (Barry and Uzuner, 2019), (Aroyehun and Gelbukh, 2019), (Bagherzadeh et al., 2018). Deep models, however, do not usually have access to outside resources, apart from word embeddings. While such models can outperform systems that are limited to look-up in gazetteer lists for task specific terms, this can only be when the terms of the test set are foreshadowed sufficiently in the training set.

Sensitivity to lexical triggers is crucial in classification, especially in the medical domain, where vocabularies are ever-growing and new specialized terms are introduced everyday. The most recent example is the term "CoVID" which was coined in late 2019.

Most language models are trained contextually. The assumption for contextualized language models is that the meaning of a word can be represented by the context in which it appears. However, the context usually is not sufficient to represent the meaning for rare specialized terms, which require large amounts of training data for coverage. In addition, highly specialized terms with very different meanings may occur in the same immediate context (see Example 1), rendering contextualized word embeddings less effective.

(1) (a) *My son was diagnosed with leukemia*
    (b) *My son was diagnosed with hydrocephalus*

The context for *hydrocephalus* and *leukemia* here is the same, and is the same for all diseases, making such contextualized word embeddings less sensitive to, for instance, the more fine-grained distinctions between *birth defects* and *cancer*. Consequently, contextualized language models often fail to make these distinctions.

Current language models have over 60M parameters[1], making fine-tuning as well as testing time-consuming and requiring large training sets.

[1]The smallest model, DistilBERT (Sanh et al., 2019), has 66M parameters, $\mathrm{RoBERTa}_{Large}$ (Liu et al., 2019) has 340M parameters.

Proceedings of the 12th International Workshop on Health Text Mining and Information Analysis, pages 38–48
April 19, 2021.

These issues motivate us to investigate widely available external knowledge sources, such as MeSH (Lipscomb, 2000), and language features in a deep architecture suitable for personal health mention detection. We show that knowledge sources, combined with light-weight word embeddings and language models such as GLoVE (Pennington et al., 2014) and ELMo (Peters et al., 2018), are strong contenders for larger models such as RoBERTa (Liu et al., 2019).

We experiment on four health-related tweet classification tasks of the ongoing SMM4H Workshop and present ablation studies to assess the contribution of different external knowledge sources. Our results suggest that the external resources tested indeed enhance performance when properly calibrated to work together. Best performance is achieved with a two layer system that adds representations of gazetteer lists and enhanced part-of-speech annotations in an encoder followed by a graph convolutional neural network (GCNN) (Kipf and Welling, 2017) representing preprocessed grammatical dependencies.

2 Related literature

Personal experiences posted on social media can give insight into the state of public health. Examination and identification of smoking behavior (Myslín et al., 2013), non-medical use of opioids (Chan et al., 2015), and identification of medication-related experiences (Jiang et al., 2018), (Jiang et al., 2019) have recently been studied on social media.

A variety of models have been proposed for the task addressed in the current paper, namely health experience mention detection for different experiences. The approaches fall into three main categories, namely statistical models with hand-crafted features, pure deep learning models, and deep models with leveraged features.

Hand-crafted features (Jiang et al., 2016) proposed a set of textual features such as count of emotion words, of unique words, of first person pronouns, of pronouns, etc., for personal health surveillance. (Jiang et al., 2019) compared different word embeddings such as GLoVE, Word2Vec (Mikolov et al., 2013), fastText, and wordRank with the features of (Jiang et al., 2016). Their word embeddings performed close to one another and considerably outperform their feature based model.

For detection of vaccination behaviour, (Joshi et al., 2018) proposed to use the count of POS tags, number of special characters (such as # and @), and count of emotion words as input features to an ensemble of SVM, logistic regression and random forest classifiers. (Joshi et al., 2018) also experimented with a pure deep-learning method (employing ULMfit (Howard and Ruder, 2018) with fine-tuning) and reported a performance close to their feature-based model, demonstrating that a model with handcrafted features is a strong contender for deep models.

To identify drug and adverse drug reaction mentions, (Saha et al., 2018) used a SVM classifier with some hand-crafted features such as the count of typed-dependency relation, drug names, and sentiment score and demonstrated success to some extent. (Çöltekin and Rama, 2018) also addressed the task using a SVM model with word and character n-grams as input features. Bag of word features (tf-idf) as well as negation, adverse reaction mentions, and drug mention were also used in (Wang et al., 2019) as input for a SVM.

Models with hand-crafted features have shown competitive performance compared to deep models on some tasks and datasets but despite the availability of many high quality resources from which features can be derived, the power of contextualized word embeddings to fill in not only lexical gaps, but also subtask specific patterns led to the investigation of deep models.

Pure deep models A majority of the proposed models for personal health detection are deep learning based. Studies such as (Xherija, 2018) and (Cortes-Tejada et al., 2019) use conventional pre-trained word embeddings, such as Word2Vec (Mikolov et al., 2013) and GLoVe. By the advent of pre-trained language models, many studies benefit from models such as ULMfit (Howard and Ruder, 2018), ELMo (Peters et al., 2018), and BERT (Devlin et al., 2019) variants (Khan et al., 2020), (Sarabadani, 2019), (Miftahutdinov et al., 2019), (Aroyehun and Gelbukh, 2019), (Babu and Eswari, 2020), (Aduragba et al., 2020).

Deep models are dependent on the representativeness of their training data for the test cases. Pretrained (often BERT-based models) are generally the top performers in current shared task competitions. The models have to be developed by highly skilled machine learning experts. The data, on the other hand, have to be collected and annotated by domain experts, if they are to be of use in health care research.

Deep models with leveraged features While many approaches to health-oriented classification of tweets use features in statistical models, there have been fewer efforts to leverage them in deep models.

For the task of adverse drug reaction mention detection (Wu et al., 2019) proposed to embed POS tags, gazetteer information, and sentiment scores, then concatenate the features to GloVe embeddings as input to a hybrid of CNN and LSTM. POS tags as well as features such as side effects, medical concepts, and first character are concatenated to word embeddings in (Vydiswaran et al., 2019). (Bagherzadeh et al., 2019) also leveraged features such as adverse mentions, POS tags, and scope of negation and modality, by concatenation to Word2Vec and GLoVE embeddings.

The limitation of input feature concatenation is that only a fixed number of annotation types can be used. Adding more gazetteer lists requires reconfiguring the network, since the number of dimensions is bounded by the number of predefined annotations. This is undesirable because addition of any annotation type has to be performed by a machine learning expert and must be followed by re-training the network from scratch.

In the following we outline a simple architecture, where the end-user (possibly an epidemiologist) can add or remove gazetteer lists and fine-tune the model without making any changes in the network settings such as hidden dimensions (and thus without requiring help from a machine learning expert).

3 Tasks

In order to demonstrate the ability of the presented approach to adapt to new domains, we compare performance on four tasks from the Social Media Mining for Health application[2] (SMM4H) shared tasks. All tasks involve detection of self-reported health mentions on Twitter.

SM18-2: Self-reported medication intake is a 3-class. Tweets which clearly express personal medication intake are considered Category 1. Tweets where the user *may* have taken some medication are Category 2. Category 3 tweets mention medication names but do not indicate personal intake (Weissenbacher et al., 2018).

(2) (Class 1):
*I took three Ibuprofens and I still got a headache crack head *cough cough**

(3) (Class 2):
since I'm constantly in pain, the only way I can go to sleep is if I take Tylenol PM

(4) (Class 3):
Will you take a Xanax and relax.

The performance is evaluated as μF score of Class 1 and Class 2.[3]

SM18-4 Vaccination behavior mention classification is a binary task where the positive class indicates the user has received or intends to receive a flu vaccine. A tweet is classified as negative if it does not contain any mentions of a vaccination or if it merely mentions vaccination (Weissenbacher et al., 2018).

(5) (Class 0):
scientists found a flu vaccine flaw, now they have to fix it

(6) (Class 1):
waiting at the pharmacy for my flu shot

SM19-1 Adverse drug reaction mention is the task of identifying mentions of side effects as the results of drug consumption (Weissenbacher et al., 2019).

(7) (Class 0):
I'm so proud of bob for taking xarelto!

(8) (Class 1):
This Vyvanse got me sweating right now and I dont even know why

SM20-5 Birth defect mention detection is a 3-way classification problem, where Category 1 tweets refer to the user's child and indicate that he/she has a birth defect. Category 2 tweets are unclear whether the tweet speaks of birth defects of the author's child. Category 3 tweets merely mention birth defects but not with respect to the author's child (Klein et al., 2020). Examples of each class are provided in Examples 9-11.

(9) (Class 1):
I had a stillbirth when I was 7 month pregnant. It was hydrocephalus.

[2] https://healthlanguageprocessing.org/smm4h-2021/

[3] for all tasks we follow the standard measure used in SMM4H competitions

(10) (Class 2):
Olivia was born with down syndrome.

(11) (Class 3):
Down's syndrome day. Please share to raise awareness.

The performance is evaluated as the μF score of Class 1 and Class 2.

A summary of the statistics of training and test data is provided in Table 1.

Table 1: Statistics of the data sets

Task	Train	Test
SM18-2	14219	3554
SM18-4	4579	1144
SM19-1	25678	4575
SM20-5	18382	4603

4 External resources

We experiment with two types of external knowledge sources for the deep learning system: (a) gazetteer lists extracted from MeSH as examples of a high quality resource developed by experts that can be used to partly define the domain of the task and (b) language features (POS, NEs, dependencies) extracted from the text with a parser and named entity recognition pipeline.

4.1 Gazetteer lists

Disease Mentions of disease are important evidence for medication intake classification, since drugs are usually consumed to treat a disease or its symptom. To identify disease mentions, we compiled a gazetteer from subtree C in MeSH (Lipscomb, 2000) which includes terms for *infections, wounds, injuries, pain, etc.*

(12) *: I've literally had a <u>headache</u> all day today and have taken four Tylenols throughout the day !*

Drug To identify drug mentions, we use the DrugBank database (Wishart et al., 2018), which includes commercial drug names as well as their scientific names.

Anatomy Body parts are often present in both birth defect and adverse drug reaction mentions. Tweets talk about a child's birth defect often specifically mentioning an affected body part. When talking about an adverse drug reaction, tweets often mention affected organs. To identify these mentions of anatomy, we extracted a gazetteer list from sub-tree A of MeSH.

ADR We use the adverse drug reaction (ADR) lexicon provided by (Nikfarjam et al., 2015) which is a collection of several lexica including SIDER (Kuhn et al., 2016), CHV (Zeng et al., 2007), COSTART,[4] and DIEGO_Lab ADR lexicon[5].

Preg For tweets that mention a pregnancy issue, rather than a birth defect, a gazetteer list of pregnancy complication terms was extracted from subtree C13.703 of MeSH.

(13) *After a <u>stillbirth</u> in 2014 for #Trisomy18, yesterday we found out we are expecting a healthy baby*

BirthDef Terms referring to birth defects vary. For instance, *Down's Syndrome* is variously referred to as *Mongolism*, *Trisomy G*, and *Trisomy 21*. Supplementing the gazetteer list potentially enables the model to make correct predictions for those instances of birth defects that have not been observed in the training data. We compiled a gazetteer list of congenital, hereditary, and neonatal diseases and abnormalities from MeSH C16.

Descendant Terms referring to children: *kid, son, daughter, baby, child, fetus, girl, boy, infant, toddler, twins*. In addition, this gazetteer includes patterns such as *one year old* described by the regular expression:

CD (-|)?(day|days|month|months|year|years)(-|)?old

where ? denotes zero or one instance of a token, | denotes alternatives, and *CD* indicates a number.

FamilyRel Family terms such as *mom, dad, mommy, daddy, mother, father, grandfather, grandmother, wife, husband, spouse, sister, brother, parent, sister in law, brother in law, cousin, niece, nephew*

Acquaintances Terms such as *neighbor, friend, colleague*, etc., from GI (Stone et al., 1966) (tagged as SocRel)

The last two gazeteers enable the system to distinguish examples of self reports (*my child*) from reports on a family member (*my cousin's kid*).

[4] http://bioportal.bioontology.org/ontologies/COSTART
[5] http://diego.asu.edu/Publications/ADRMine.html

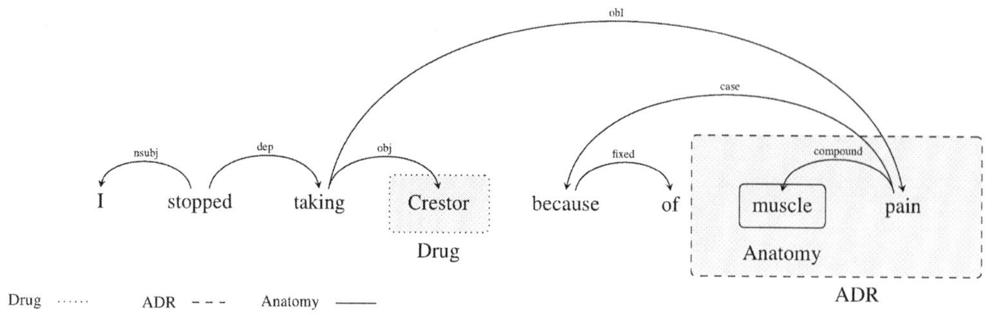

Figure 1: Dependency parse for *I stopped taking Crestor because of muscle Pain*, with gazetteer annotations

4.2 Named entities

Proper names and names of organizations are extracted using the ANNIE module.

Names People often mention the name of their child when talking about a personal experience.

(14) *My Kristin such a blessing from GOD - Kids with Down Syndrome*

Organization The presence of an organization mention often indicates that a tweet is talking in a general sense and not relating a personal experience.

(15) *Ohio Senate Says 'No' to Abortion Based on Down Syndrome Diagnosis*

Gazetteer lists have two main advantages. First, they enable the model to extend its vocabulary. Second, they determine the position of a certain annotation type in a sentence, which becomes important when coupled with dependency relations, as illustrated in Figure 1. The mapping[6] of gazetteer lists to tasks is provided in Table 2.

Table 2: The set of gazetteer lists used for each task, subsumed under the label *Gaz*

Task	Gazetteer set
SM18-2	Drug, Disease
SM18-4	Descendant, FamilyRel, Acquaintance
SM19-1	Drug, Disease, ADR, Anatomy, Descendant, FamilyRel, Acquaintance
SM20-5	BirthDef, Preg, Anatomy, Descendant, FamilyRel

4.3 POS tags

Part-of-speech tags are the most widely used linguistic feature and are available from many standard NLP environments. POS tags provide useful information such as types of pronouns and tense for verbs, all important clues for the detection of a personal experience. POS tags have been used in the literature for classifying personal and impersonal sentences (Li et al., 2010).

Since our tasks focus intensely on first person reports, we replace the single tag for pronouns *PRP* with three tags *PRP1, PRP2, PRP3*, reserved for first, second, and third person pronouns. Likewise, the reflexive pronoun tag *PRP$* is replaced by *PRP$1, PRP$2, and PRP$3* for first, second, and third person possessive pronouns. In our experiments we compare the standard Penn Treebank tag set (denoted by POS1) to this extended POS tag set (denoted by POS2).

While POS-tag information is partially encoded in word embeddings, our ablation shows that explicit encoding leads to performance increase and that POS2 is part of our best performing model.

4.4 Dependency parse

Dependency graphs provide syntactic knowledge as well as shallow semantic information. An example of a dependency graph for the ADR task together with gazetteer annotations is provided in Figure 1.

Some dependency relations are indicative of personal experience mentions. For instance, drug or birth defect mentions occur more likely as direct objects. Self-reports mostly use first person pronouns in subject position.

We use the Stanford parser (Klein and Manning, 2003) to determine dependency relations.

[6]Note that in Table 4, the label *Gaz* refers to the respective gazetteer set

	h_1^1	h_2^1	h_3^1	h_4^1	h_5^1	h_6^1	h_7^1	h_8^1	h_9^1
	=	=	=	=	=	=	=	=	=
POS Embedding	P_{NNP}	P_{VBZ}	P_{NNP}	P_{VBG}	P_{PRP3}	P_{TO}	P_{VB}	P_{NN}	P_{NN}
	+	+	+	+	+	+	+	+	+
Gazetteer Embedding	G_{Name}	**0**	$G_{BirthDef}$	**0**	**0**	**0**	**0**	$G_{Anatomy}$	**0**
	+	+	+	+	+	+	+	+	+
Token Embedding	E_{Milo}	E_{has}	$E_{Hydrocephalus}$	$E_{causing}$	E_{him}	E_{to}	E_{need}	E_{brain}	E_{shunt}
Input	Milo	has	Hydrocephalus	causing	him	to	need	brain	shunt
time-step	1	2	3	4	5	6	7	8	9

Figure 2: Additive annotation embedding (Layer1)

5 Proposed model

We developed a multi-layer system which includes four layers, namely: embeddings, self-attention, GCNN, and classification.

Layer1: Embeddings We combine traditional word embeddings with POS embeddings and our gazetteer embeddings additively.

- Tokens are embedded by GLoVE,[7] ELMo,[8], pretrained RoBERTa[9] (Liu et al., 2019), or BioBERT (Lee et al., 2020).
- We pretrain POS embeddings using Word2Vec. Our approach is to apply Word2Vec on POS tags instead of tokens. The embeddings are trained using the Gensim package (Rehurek and Sojka, 2010) with a window size of $w = 5$. The pretraining is performed on training data of all task introduced in Section 3. The resulting embeddings are used to initialize a POS embedding matrix $P \in \mathbb{R}^{\phi \times d_{emb}}$, where ϕ is the number of distinct POS tags, and d_{emb} is the dimensionality of the word embeddings. The POS embeddings are fine-tuned during the training for the main classification task.
- A gazetteer annotation x is embedded through a vector $G_x \in \mathbb{R}^{d_{emb}}$ which is a learnable parameter and is updated during the training of the main classification task.
- Inspired by BERT (Devlin et al., 2019) segment embeddings, we add POS embeddings and gazetteer embeddings to token embeddings. We call this scheme, additive annotations. Figure 2 shows that for time step 3, the vectors $E_{Hydrocephalus}$, $G_{BirthDef}$, and P_{NNP} are added to form h_3^1, the aggregate for time-step 3 in layer 1.

The additive approach enables the model to encode as many as gazetteer annotations, without introducing new dimensions to the token representations (in contrast to concatenative approaches). After the training, one can easily introduce a new gazetteer annotation y by adding a learnable vector G_y to the model parameters, and only fine-tune the model, without making any changes to hidden dimensions.

Layer2: Self-attention encoder We use a self-attention encoder (the encoder part of the Transformer) as first layer (Vaswani et al., 2017). The encoder at layer 2 gets the representations h_i^1 and outputs representations h_i^2. The number of heads in the multi-head attention is $n_{heads} = 4$ and the dimensionality of the feed-forward layer is $d_{FF} = 1024$.

Layer3: Graph CNN We use a graph convolutional network (GCNN) (Kipf and Welling, 2017) to encode the dependency graph following (Marcheggiani and Titov, 2017). In GCCN, each token is represented based on its adjacent tokens in a dependency parse using:

$$h_i^3 = ReLU\left(\sum_{j \in \mathcal{N}(i)} W_{L(i,j)} h_j^2 + b\right) \quad (1)$$

where $\mathcal{N}(i)$ is the set of tokens adjacent to token i and $L(i,j)$ is the label of the arc from token j to token i. Note that the network is not tied, i.e. $W_{L(i,j)}$ depends on the arc labels. GCNN receives h_i^2 and outputs token-wise representations h_i^3.

Layer4: Pooling and classification For the vector representation of the tweet, attention (Bahdanau et al., 2015) is calculated from importance scores:

$$e_i = w_{att}^T h_i^3 \quad (2)$$

[7]Twitter ($d_{emb} = 200$)
[8]Original ($d_{emb} = 1024$)
[9]Large ($d_{emb} = 1024$)

Table 3: The set of hyper-parameters used for each task

	Embedding	Epoch	lr
SM18-2	GLoVE	5	.1e-3
	ELMo	6	.5e-4
	RoBERTa / BioBERT	10	.5e-5
SM18-4	GLoVE	4	.1e-3
	ELMo	6	.1e-3
	RoBERTa / BioBERT	6	.1e-4
SM19-1	GLoVE	6	.1e-3
	ELMo	8	.1e-4
	RoBERTa / BioBERT	10	.1e-5
SM20-5	GLoVE	4	.1e-4
	ELMo	6	.1e-4
	RoBERTa / BioBERT	8	.5e-5

using a latent context vector w_{att}, and then normalizing the scores using softmax:

$$\alpha_i = \frac{exp(e_i)}{\sum_j exp(e_j)} \tag{3}$$

The normalized scores are then used for a weighted sum $H = \sum_i \alpha_i * h_i^3$. The final vector H is used as input to a linear layer for the classification.

The proposed model is implemented using the PyTorch library (Paszke et al., 2017). Cross-entropy is used to calculate the network loss and the model is optimized using the Adam optimizer (Kingma and Ba, 2015). Table 3 details the hyper-parameters used for each task.

6 Numerical results

We evaluate the proposed model using a set of ablation studies. The SM19-1 and SM20-5 tasks are evaluated on the official test data. For SM18-2 and SM18-4, official test data is not available, therefore we replicate the state-of-the art systems and perform evaluation on a hold-out set from the original training data.

Table 4 shows that all tasks benefit moderately from POS features with the extended POS tagset POS2 outperforming the standard Penn Treebank tagset POS1. POS features increase performance for GLoVE and ELMO more than for RoBERTa or BioBERT. Dependency information Dep, similarly, yields consistent small improvements. Note, however, the asymmetrically stronger improvements in precision, especially for RoBERTa and BioBERT models. Combining POS and Dep results in another consistent small improvement, showing that the features effectively interoperate.

The gazetteer lists and named-entity features provide considerable improvements for all tasks except for SM18-4 with marginal improvements. Note that SM18-4 is the vaccination behaviour prediction task, specific to flu. This result is to be expected: it requires identifying self reports, but the trigger terms for the flu domain consisted only of *flu*, making gazetteers ineffective. The tasks with more diverse vocabularies show greater impact of gazetteer lists.

Combining grammatical and gazetteer features robustly yields best results. Interestingly, adding knowledge resources to a lighter language model approaches performance of a larger model. For instance, GloVe with all resources outperforms ELMo without resources for all tasks and even approaches RoBERTa or BioBERT without resources.

We also observe that while our system configurations using RoBERTa or BioBERT reported in Table 4 beat the SOTA reported in competition in F1 and precision, our recall only exceeds SOTA for SM18-4 and SM18-4. We interpret this as a strong point of our system: in health-related applications, precision often outweighs recall. For system development, increasing recall is usually easier and this paper limits itself to gazetteer lists and linguistic features for domain adaptation to show their potential and limits. This leaves room for further error-driven domain adaptation.

7 Case-study

To examine the effects of knowledge sources, we probe the attention importance scores at Layer4 (Equation 2). The scores demonstrate how the the model attends to different tokens. We probe the scores in two cases, with and without gazetteer lists. Figure 3 and Figure 4 demonstrate visualizations of the attention scores for two samples from SM20-5 and SM19-1 tasks respectively. Higher attention scores are indicated with darker gray color.

The model of Figure 3a uses no gazetteer lists. The model partially attends to the birth defect mention *Trisomy18* and pays no attention to the pregnancy issue *StillBirth*. The model, however, properly attends to the personal pronouns *I* and lexical triggers such as *baby* and *birth*. On the other hand, when the model is given BirthDef and Pregnancy gazetteers, the model puts more attention on *StillBirth* as evidence for a birth defect, leading to a more certain prediction.

A similar pattern is observed in Figure 4. Supply-

Table 4: Ablation of grammatical features and gazetteer lists

	Features	*SM18-2*			*SM18-4*			*SM19-1*			*SM20-5*		
		μP	μR	μF	P	R	F	P	R	F	μP	μR	μF
GLoVE	None	.63	.64	.63	.78	.76	.77	.50	.51	.50	.54	.53	.54
	POS1	.64	.67	.66	.79	.78	.78	.51	.55	.53	.53	.57	.55
	POS2	.66	.67	.66	.79	.81	.80	.53	.55	.54	.53	.59	.56
	Dep	.68	.65	.67	.83	.75	.79	.56	.53	.54	.58	.58	.58
	POS1, Dep	.70	.66	.68	.81	.78	.79	.57	.55	.56	.60	.59	.60
	POS2, Dep	.70	.68	.69	.82	.78	.80	.58	.58	.58	.61	.60	.60
	Gaz, Name, Org	.69	.72	.71	.79	.77	.78	.55	.56	.55	.60	.58	.59
	Gaz, Name, Org, Pos2, Dep	.73	.71	.72	.83	.78	.81	.59	.60	.60	.65	.60	.62
ELMo	None	.71	.69	.70	.80	.78	.79	.54	.53	.54	.64	.60	.62
	POS1	.73	.68	.70	.79	.81	.80	.54	.54	.54	.66	.61	.63
	POS2	.72	.70	.71	.80	.81	.81	.53	.57	.55	.65	.63	.64
	Dep	.75	.68	.71	.86	.78	.82	.58	.56	.57	.68	.61	.65
	POS1, Dep	.74	.71	.72	.85	.80	.83	.57	.59	.58	.70	.62	.66
	POS2, Dep	.75	.71	.73	.86	.81	.84	.59	.60	.60	.71	.62	.67
	Gaz, Name, Org	.72	.76	.74	.80	.80	.80	.58	.56	.57	.69	.66	.68
	Gaz, Name, Org, Pos2, Dep	.77	.73	.75	.85	.83	.84	.61	.61	.61	.71	.65	.69
RoBERTa	None	.71	.74	.73	.87	.82	.85	.62	.58	.60	.68	.62	.65
	POS1	.69	.76	.73	.85	.86	.85	.61	.60	.60	.72	.64	.69
	POS2	.71	.75	.74	.87	.87	.87	.62	.60	.61	.74	.65	.70
	Dep	.76	.73	.74	.89	.86	.87	.64	.59	.61	.74	.60	.67
	POS1, Dep	.75	.75	.75	.87	.88	.87	.62	.62	.62	.75	.65	.70
	POS2, Dep	.76	.77	.76	.88	.88	.88	.63	.62	.62	.76	.65	.71
	Gaz, Name, Org	.73	.76	.75	.89	.83	.86	.65	.60	.63	.76	.65	.71
	Gaz, Name, Org, Pos2, Dep	.77	.78	.77	.90	.89	.89	.69	.62	.66	.79	.67	.73
BioBERT	None	.72	.76	.74	.85	.83	.84	.61	.63	.62	.67	.65	.66
	POS1	.73	.75	.74	.85	.86	.85	.63	.63	.63	.69	.67	.68
	POS2	.73	.75	.74	.86	.87	.86	.64	.63	.63	.69	.68	.69
	Dep	.78	.72	.75	.88	.88	.88	.66	.63	.65	.74	.62	.68
	POS1, Dep	.78	.73	.75	.88	.89	.88	.65	.65	.65	.75	.67	.71
	POS2, Dep	.78	.74	.76	.88	.89	.89	.65	.66	.65	.76	.68	.72
	Gaz, Name, Org	.74	.77	.76	.88	.84	.86	.64	.62	.63	.76	.68	.72
	Gaz, Name, Org, Pos2, Dep	.77	.77	.77	.90	.89	.89	.68	.65	.67	.79	.68	.73
	SOTA:	.63	.77	.70	.82	.79	.80	.60	.68	.64	.65	.73	.69
		(Xherija, 2018)			(Joshi et al., 2018)			(Chen et al., 2019)			(Bai and Zhou, 2020)		

ing the model with the Anatomy and Drug gazetteer leads the model to pay more attention to drug mentions and mentions of affected body parts.

8 Conclusions

This paper demonstrates the effectiveness of using gazetteer lists from high-quality sources, standard named entity categories and part-of-speech embeddings with a self-attention encoder and a GCNN encoding grammatical dependencies. The architecture supports precision oriented domain adaptation from widely available, high-quality resources (i.e. MeSH). Adaptation with new gazetteer lists using additive annotation sidesteps the need to reconfigure or retrain the neural networks

The experiments confirm that quality external resources can offset the lower parameter space of light-weight word embedding/language models, such as GLoVE and ELMo. At the same time, these resources effectively combine with RoBERTa for best performance. The stronger improvements in precision are especially promising for health applications.

The comparative results on different tasks and different domains proves that this extensible architecture is well-suited for actual use in the wild on domains and tasks, for which experts know to supply high-quality terminology resources.

Acknowledgement

We would like to thank the anonymous reviewers for their valuable and constructive comments.

Its been a year since I found out I 'd be giving birth to a sleeping baby ! # StillBirth ! #Loss # Trisomy18

a) Without Birth Defect and Pregnancy gazetteers

Its been a year since I found out I 'd be giving birth to a sleeping baby ! # StillBirth #Loss # Trisomy18

b) With Birth Defect and Pregnancy gazetteers

Figure 3: Visualization of attention scores for a sample from SM20-5

after taking olanzapine i wake up and feel like i am in a straight jacket because my muscles feel stiff

a) Without Drug and Anatomy gazetteers

after taking olanzapine i wake up and feel like i am in a straight jacket because my muscles feel stiff

b) With Drug and Anatomy gazetteers

Figure 4: Visualization of attention scores for a sample from SM19-1

References

Olanrewaju Tahir Aduragba, Jialin Yu, Gautham Senthilnathan, and Alexandra Crsitea. 2020. Sentence Contextual Encoder with BERT and BiLSTM for Automatic Classification with imbalanced medication tweets. In *SMM4H 2020.*

Segun Taofeek Aroyehun and Alexander Gelbukh. 2019. Detection of Adverse Drug Reaction in Tweets Using a Combination of Heterogeneous Word Embeddings. In *SMM4H 2019.*

Yandrapati Prakash Babu and Rajagopal Eswari. 2020. Identification of Medication Tweets Using Domain-specific Pre-trained Language Models. In *SMM4H 2020.*

Parsa Bagherzadeh, Nadia Sheikh, and Sabine Bergler. 2018. CLaC at SMM4H task 1, 2, and 4. In *SMM4H 2018.*

Parsa Bagherzadeh, Nadia Sheikh, and Sabine Bergler. 2019. Adverse Drug Effect and Personalized Health Mentions, CLaC at SMM4H 2019, Tasks 1 and 4. In *SMM4H 2019.*

Dzmitry Bahdanau, Kyunghyun Cho, and Yoshua Bengio. 2015. Neural machine translation by jointly learning to align and translate. In *Proceedings of the 3rd International Conference on Learning Representations, ICLR'15.*

Yang Bai and Xiaobing Zhou. 2020. Automatic Detecting for Health-related Twitter Data with BioBERT. In *SMM4H 2020.*

Paul Barry and Ozlem Uzuner. 2019. Deep Learning for identification of adverse effect mentions in Twitter data. In *SMM4H 2019.*

Brian Chan, Andrea Lopez, and Urmimala Sarkar. 2015. The canary in the coal mine tweets: social media reveals public perceptions of non-medical use of opioids. *PloS one*, 10(8).

Lauren E Charles-Smith, Tera L Reynolds, Mark A Cameron, Mike Conway, Eric HY Lau, Jennifer M Olsen, Julie A Pavlin, Mika Shigematsu, Laura C Streichert, Katie J Suda, et al. 2015. Using social media for actionable disease surveillance and outbreak management: a systematic literature review. *PloS one*, 10(10).

Shuai Chen, Yuanhang Huang, Xiaowei Huang, Haoming Qin, Jun Yan, and Buzhou Tang. 2019. HITSZ-ICRC: A report for SMM4H shared task 2019-automatic classification and extraction of adverse effect mentions in tweets. In *SMM4H 2019.*

Çağrı Çöltekin and Taraka Rama. 2018. Drug-use Identification from Tweets with Word and Character N-grams. In *SMM4H 2018.*

Javier Cortes-Tejada, Juan Martinez-Romo, and Lourdes Araujo. 2019. NLP@ UNED at SMM4H 2019: Neural Networks Applied to Automatic Classifications of Adverse Effects Mentions in Tweets. In *SMM4H 2019.*

Jacob Devlin, Ming-Wei Chang, Kenton Lee, and Kristina N. Toutanova. 2019. BERT: Pre-training of Deep Bidirectional Transformers for Language Understanding. In *Proceedings of NAACL-HLT 2019.*

Su Golder, Gill Norman, and Yoon K Loke. 2015. Systematic review on the prevalence, frequency and comparative value of adverse events data in social media. *British journal of clinical pharmacology*, 80(4).

Jeremy Howard and Sebastian Ruder. 2018. Universal Language Model Fine-tuning for Text Classification. In *ACL 2108.*

Keyuan Jiang, Ricardo Calix, and Matrika Gupta. 2016. Construction of a personal experience tweet corpus for health surveillance. In *BioNLP.*

Keyuan Jiang, Shichao Feng, Ricardo A Calix, and Gordon R Bernard. 2019. Assessment of word embedding techniques for identification of personal experience tweets pertaining to medication uses. In *International Workshop on Health Intelligence.*

Keyuan Jiang, Shichao Feng, Qunhao Song, Ricardo A Calix, Matrika Gupta, and Gordon R Bernard. 2018. Identifying tweets of personal health experience through word embedding and LSTM neural network. *BMC Bioinformatics*, 19(8).

Aditya Joshi, Xiang Dai, Sarvnaz Karimi, Ross Sparks, Cecile Paris, and C Raina MacIntyre. 2018. Shot or not: Comparison of NLP approaches for vaccination behaviour detection. In *SMM4H 2018*.

Pervaiz Iqbal Khan, Imran Razzak, Andreas Dengel, and Sheraz Ahmed. 2020. Improving Personal Health Mention Detection on Twitter Using Permutation Based Word Representation Learning. In *NeurIPS 2020*. Springer.

Diederik P Kingma and Jimmy Lei Ba. 2015. Adam: A method for stochastic optimization. In *Proceedings of the 3rd International Conference on Learning Representations, ICLR'15*.

Thomas. N. Kipf and Max Welling. 2017. Semi-supervised classification with graph convolutional networks. In *ICLR'17*.

Ari Z. Klein, Ivan Flores, Arjun Magge, Anne-Lyse Minard, Karen O'Connor, Abeed Sarker, Elena Tutubalina, Davy Weissenbacher, and Graciela Gonzalez-Hernandez. 2020. Overview of the Fifth Social Media Mining for Health applications (SMM4H) shared tasks at COLING 2020. In *SMM4H 2020*.

Dan Klein and Christopher D. Manning. 2003. Accurate Unlexicalized Parsing. In *ACL 2003*.

Michael Kuhn, Ivica Letunic, Lars Juhl Jensen, and Peer Bork. 2016. The SIDER database of drugs and side effects. *Nucleic acids research*, 44(D1).

Jinhyuk Lee, Wonjin Yoon, Sungdong Kim, Donghyeon Kim, Sunkyu Kim, Chan Ho So, and Jaewoo Kang. 2020. Biobert: a pre-trained biomedical language representation model for biomedical text mining. *Bioinformatics*, 36(4):1234–1240.

Shoushan Li, Chu-Ren Huang, Guodong Zhou, and Sophia Yat Mei Lee. 2010. Employing personal/impersonal views in supervised and semi-supervised sentiment classification. In *ACL 2010*.

Carolyn E Lipscomb. 2000. Medical subject headings (MeSH). *Bulletin of the Medical Library Association*, 88(3).

Yinhan Liu, Myle Ott, Naman Goyal, Jingfei Du, Mandar Joshi, Danqi Chen, Omer Levy, Mike Lewis, Luke Zettlemoyer, and Veselin Stoyanov. 2019. RoBERTa: A robustly optimized BERT pretraining approach. *arXiv preprint arXiv:1907.11692*.

Diego Marcheggiani and Ivan Titov. 2017. Encoding sentences with graph convolutional networks for semantic role labeling. In *EMNLP 2017*.

Zulfat Miftahutdinov, Ilseyar Alimova, and Elena Tutubalina. 2019. KFU NLP team at SMM4H 2019 tasks: Want to extract adverse drugs reactions from tweets? BERT to the rescue. In *SMM4H 2019*.

Tomas Mikolov, Ilya Sutskever, Kai Chen, Greg S. Corrado, and Jeff Dean. 2013. Distributed representations of words and phrases and their compositionality. In *Advances in Neural Information Processing Systems*.

Mark Myslín, Shu-Hong Zhu, Wendy Chapman, and Mike Conway. 2013. Using Twitter to examine smoking behavior and perceptions of emerging tobacco products. *Journal of medical Internet research*, 15(8).

Azadeh Nikfarjam, Abeed Sarker, Karen O'Connor, R Ginn Rachel, and Garciela Gonzalez. 2015. Pharmacovigilance from social media: mining adverse drug reaction mentions using sequence labeling with word embedding cluster features. *Journal of the American Medical Informatics Association*, 22(3).

Adam Paszke, Sam Gross, Soumith Chintala, Gregory Chanan, Edward Yang, Zachary DeVito, Zeming Lin, Alban Desmaison, Luca Antiga, and Adam Lerer. 2017. Automatic differentiation in PyTorch. In *NIPS 2017*.

Jeffrey Pennington, Richard Socher, and Christopher D. Manning. 2014. Glove: Global Vectors for Word Representation. In *EMNLP 2014*.

Matthew E. Peters, Mark Neumann, Mohit Iyyer, Matt Gardner, Christopher Clark, Kenton Lee, and Luke Zettlemoyer. 2018. Deep contextualized word representations. In *Proceedings of NAACL*.

Radim Rehurek and Petr Sojka. 2010. Software framework for topic modelling with large corpora. In *In Proceedings of the LREC 2010 Workshop on New Challenges for NLP Frameworks*.

Dimitrios Rousidis, Paraskevas Koukaras, and Christos Tjortjis. 2020. Social media prediction: a literature review. *Multimedia Tools and Applications*, 79(9).

Rupsa Saha, Abir Naskar, Tirthankar Dasgupta, and Lipika Dey. 2018. Leveraging web based evidence gathering for drug information identification from tweets. In *SMM4H 2018*.

Victor Sanh, Lysandre Debut, Julien Chaumond, and Thomas Wolf. 2019. DistilBERT, a distilled version of BERT: smaller, faster, cheaper and lighter. In *NeurIPS 2019*.

Sarah Sarabadani. 2019. Detection of adverse drug reaction mentions in tweets using ELMo. In *SMM4H 2019*.

Philip J Stone, Dexter C Dunphy, and Marshall S Smith. 1966. *The General Inquirer: A computer approach to content analysis*. MIT press.

Ashish Vaswani, Noam Shazeer, Niki Parmar, Jakob Uszkoreit, Llion Jones, Aidan N Gomez, Łukasz Kaiser, and Illia Polosukhin. 2017. Attention is all you need. In *NIPS 2017*.

VG Vinod Vydiswaran, Grace Ganzel, Bryan Romas, Deahan Yu, Amy Austin, Neha Bhomia, Socheatha Chan, Stephanie Hall, Van Le, Aaron Miller, et al. 2019. Towards text processing pipelines to identify adverse drug events-related tweets: University of Michigan @ SMM4H 2019 Task 1. In *SMM4H 2019*.

Chen-Kai Wang, Hong-Jie Dai, and Bo-Hung Wang. 2019. BIGODM System in the Social Media Mining for Health Applications Shared Task 2019. In *SMM4H 2019*.

Davy Weissenbacher, Abeed Sarker, Arjun Magge, Ashlynn Daughton, Karen O'Connor, Michael J. Paul, and Graciela Gonzalez-Hernandez. 2019. Overview of the fourth Social Media Mining for Health (SMM4H) shared tasks at ACL 2019. In *SMM4H 2019*.

Davy Weissenbacher, Abeed Sarker, Michael J. Paul, and Graciela Gonzalez-Hernandez. 2018. Overview of the third Social Media Mining for Health (SMM4H) Shared Tasks at EMNLP 2018. In *SMM4H 2018*.

David S Wishart, Yannick D Feunang, An C Guo, Elvis J Lo, Ana Marcu, Jason R Grant, Tanvir Sajed, Daniel Johnson, Carin Li, Zinat Sayeeda, et al. 2018. DrugBank 5.0: a major update to the DrugBank database for 2018. *Nucleic acids research*, 46(D1).

Chuhan Wu, Fangzhao Wu, Zhigang Yuan, Junxin Liu, Yongfeng Huang, and Xing Xie. 2019. MSA: Jointly detecting drug name and adverse drug reaction mentioning tweets with multi-head self-attention. In *Twelfth ACM International Conference on Web Search and Data Mining*.

Orest Xherija. 2018. Classification of medication-related tweets using stacked bidirectional LSTMs with context-aware attention. In *SMM4H 2018*.

Qing Zeng, Tony Tse, Guy Divita, Alla Keselman, Jonathan Crowell, Allen Browne, Sergey Goryachev, and Long Ngo. 2007. Term identification methods for consumer health vocabulary development. *Journal of medical Internet research*, 9(1).

Integrating Higher-Level Semantics into Robust Biomedical Name Representations

Pieter Fivez
CLiPS Research Centre
University of Antwerp
pieter.fivez@uantwerpen.be

Simon Šuster
Faculty of Engineering and Information Technology
University of Melbourne
simon.suster@unimelb.edu.au

Walter Daelemans
CLiPS Research Centre
University of Antwerp
walter.daelemans@uantwerpen.be

Abstract

Neural encoders of biomedical names are typically considered robust if representations can be effectively exploited for various downstream NLP tasks. To achieve this, encoders need to model domain-specific biomedical semantics while rivaling the universal applicability of pretrained self-supervised representations. Previous work on robust representations has focused on learning low-level distinctions between names of fine-grained biomedical concepts. These fine-grained concepts can also be clustered together to reflect higher-level, more general semantic distinctions, such as grouping the names *nettle sting* and *tick-borne fever* together under the description *puncture wound of skin*. It has not yet been empirically confirmed that training biomedical name encoders on fine-grained distinctions automatically leads to bottom-up encoding of such higher-level semantics. In this paper, we show that this bottom-up effect exists, but that it is still relatively limited. As a solution, we propose a scalable multi-task training regime for biomedical name encoders which can also learn robust representations using only higher-level semantic classes. These representations can generalise both bottom-up as well as top-down among various semantic hierarchies. Moreover, we show how they can be used out-of-the-box for improved unsupervised detection of hypernyms, while retaining robust performance on various semantic relatedness benchmarks. Our code is open-source and can be found at www.github.com/clips/higherlevelsemantics.

1 Introduction

Recent work on representation learning for biomedical names has mainly involved the training of neural encoder architectures such as LSTMs (Kartsaklis et al., 2018) or Transformers (Sung et al., 2020; Kalyan and Sangeetha, 2020) to finetune name representations for biomedical normalization tasks. Such representations are often tailored towards normalization tasks (e.g. linking names to corresponding concept identifiers), without providing explicit guarantees about their transferability to other use contexts and applications. As a solution for this issue, the Biomedical Name Encoder (BNE) model (Phan et al., 2019) has been proposed as a comprehensive framework for robust and transferable representations.

According to this framework, the robustness of biomedical name representations is characterized along three dimensions. Firstly, semantic similarity between names should be reflected by their closeness in the embedding space. Secondly, the variety of textual contexts in which a name appears should be somehow represented in the encoding. Lastly, a name embedding should be sufficiently close to a pretrained prototypical representation of its conceptual meaning, e.g. a representation of its corresponding concept identifier from a biomedical ontology.

Such a multi-task model can be effectively trained using synonym sets extracted from ontologies such as the UMLS or SNOMED-CT. However, these synonym sets typically reflect only fine-grained distinctions between the lowest-level concepts from ontologies. If robust name representations should truly reflect semantic similarity in general, then the assumption is being made that training on such fine-grained synonym sets learns biomedical semantics in a bottom-up way, expecting names of lower-level concepts to spontaneously form relevant higher-level clusters.

Proceedings of the 12th International Workshop on Health Text Mining and Information Analysis, pages 49–58

Level 1	**C0564444** *wound of skin*			
Level 2	**C0561369** *puncture wound of skin*			
Level 3	**C0561546** *bite wound*		**C0576723** *sting of skin*	
Level 4	**C1302713** *animal bite wound*	**C0275134** *poisoning due to lizard venom*	**C0576722** *animal sting*	**C0576724** *plant sting*
Example name	**tick-borne fever**	**poisoning caused by gila monster venom**	**poisoning by bombus**	**nettle sting**

Table 1: Examples of how names from the SNOMED-CT ontology can be grouped into larger classes using parent concepts in the ontological graph. This allows us to investigate higher-level semantic relations, such as grouping *poisoning by bombus* and *nettle sting* under the concept of *sting of skin*, or e.g. grouping them together with *tick-borne fever* under *puncture wound of skin*.

However, such assumptions have not yet been empirically validated, for instance by showing that an encoder not only learns the differences between names such as *nettle sting* and *tick-borne fever*, but also simultaneously learns that they can be grouped together under the more general description *puncture wound of skin*. Moreover, research on representation learning and hierarchical classification for e.g. computer vision has indicated that neural models can leverage substantially different discriminative information for higher, more general levels of categorization than for more fine-grained lower levels (Hase et al., 2019). Such hierarchical differences can be exploited to generalize from higher to lower levels (Guo et al., 2017; Taherkhani et al., 2019), but they can also be difficult to integrate consistently into a single neural model (Wu et al., 2019).

In this paper, we investigate to what extent robust biomedical name representations can encode higher-level semantics while retaining relevant lower-level fine-grained information as well. To address this research question, we group synonym sets under increasingly coarse-grained semantic categories, using parent-child relations in the ontological graph. Table 1 gives an example of how names from the SNOMED-CT ontology can be grouped into larger classes. Such a hierarchy can be used to train and test a variety of semantic relations between names. For instance, a model might be able to encode that the names *poisoning by bombus* and *nettle sting* can be both described as *sting of skin*, but fail to represent their similarity to *poisoning caused by gila monster venom* as a *puncture wound of skin*. We believe that an evaluation of this nature is a crucial step towards achieving truly robust biomedical name representations, since it clearly requires more semantic inference from the encoder than merely resolving synonyms.

Apart from introducing this evaluation to the field of biomedical NLP, we also show that we can effectively adapt the BNE framework (Phan et al., 2019) to be trained using such large higher-level semantic classes. Most importantly, we replace the BiLSTM (Graves and Schmidhuber, 2005) encoder architecture of the BNE model with a lightweight Deep Averaging Network (DAN) (Iyyer et al., 2015). This allows us to easily scale to large amounts of training data, caused by the explosive amount of possible pairwise combinations between semantically similar names as classes grow larger.

Training on higher-level classes involves additional challenges such as handling imbalanced data distributions as well as implicit hierarchical and semantic differences among names grouped under the same class. Our aim is not to tailor the proposed approach to such artefacts. Rather, the main contribution of this paper is to show that our simple modification of the BNE model is generally applicable to a range of coarse-grained biomedical categorizations, without any finetuning apart from the size of the DAN encoder. As of such, it can be used as a low-cost but effective benchmark for future models that are more specialized.

Our experimental results for hierarchical SNOMED-CT data show that our DAN model improves semantic similarity ranking both in a bottom-up as well as top-down manner along various hierarchies. Interestingly, this observation holds even when we train on a few dozens of very broad categories. We also apply extrinsic evaluations to investigate the transferability of our DAN model. Firstly, we validate the robustness of higher-level representations on semantic relatedness benchmarks. Secondly, we perform unsupervised detection of SNOMED-CT hypernym disorder names which were not observed during training. For this task, our DAN model scores substantially better than the publicly released pretrained BNE model, which was trained on a large amount of fine-grained disorder concepts from SNOMED-CT using an elaborate BiLSTM architecture. These results provide tangible evidence that training name representations on large coarse-grained categories can help to encode exploitable higher-level semantics.

2 Related work

While context-dependent self-supervised representations usually outperform other text representations on a variety of BioNLP problems, such as semantic similarity and question answering, there is no single embedding model for biomedical and clinical texts that is consistently superior and thus can serve as a generally suitable bio-encoder (Tawfik and Spruit, 2020). To this date, the BNE model by Phan et al. (2019) is the most prominent attempt at developing a supervised resource for encoding biomedical names. It uses a multi-task training regime in which it combines objectives from different aspects of deep representation learning, such as a contrastive loss (Le-Khac et al., 2020), conceptual grounding (see e.g. (Kartsaklis et al., 2018)), and explicit regularization of the learned representations (e.g. used by Vulić and Mrkšić (2018)). Our modifications to the original BNE model are informed by such literature.

Our application of a Deep Averaging Network (DAN) (Iyyer et al., 2015) is inspired by a recent subfield of NLP research which has emphasized the effectiveness of random encoders (Wieting and Kiela, 2019) and simple pooling mechanisms of word embeddings. The fastText encoder which we use as a baseline and as input for the DAN is an example of a Simple Word-Embedding-based Model (SWEM) with average pooling (Shen et al., 2018).

3 Encoding model

3.1 Encoder architecture

Our encoder is a Deep Averaging Network (DAN) (Iyyer et al., 2015) which extracts a fixed-size representation for an input name n:

$$\begin{aligned} u_n &= \frac{1}{|N_t|}\sum_{t\in N_t} u_t \\ f(n) &= enc(u_n) \end{aligned} \tag{1}$$

where N_t is the bag of tokens from a name, u_t is a pretrained word embedding of a token, u_n is a name embedding created by averaging all the pretrained word embeddings of all tokens, and enc is a feedforward neural network with Rectified Linear Unit (ReLU) as non-linear activation function. As pretrained word embeddings we use 300-dimensional fastText (Bojanowski et al., 2017) representations which we train on 76M sentences of preprocessed MEDLINE articles released by Hakala et al. (2016). This fastText model also allows for constructing word embeddings for out-of-vocabulary tokens by composing character n-gram embeddings.

3.2 Training objectives

Our proposed approach is a simple modification of the multi-task training regime of the BNE model. We use cosine distance as distance function d for all three training objectives.

Semantic similarity The *semantic similarity* objective is a generalization from the synonym similarity objective of the BNE model to any level of relevant semantic similarity. To enforce embedding similarity between names that are semantically related, we use a siamese triplet loss (Chechik et al., 2010). This loss forces the encoding of a biomedical name $f(n)$ to be closer to the encoding of a semantically similar name $f(n_{pos})$ than that of an encoded negative sample name $f(n_{neg})$, within a specified (possibly tuned) margin:

$$\begin{aligned} pos &= d(f(n), f(n_{pos})) \\ neg &= d(f(n), f(n_{neg})) \\ L_{sem} &= max(pos - neg + margin, 0) \end{aligned} \tag{2}$$

To select negative names during training we apply distance-weighted negative sampling (Wu et al., 2017) over all training names, since this has been proven more effective than hard or random negative sampling.

Contextual meaningfulness The *contextual meaningfulness* objective forces the encoding of a biomedical name to be similar to its local contexts. The summary of these local contexts is approximated by taking the pretrained embedding representation u_n of the name:

$$L_{cont} = d(f(n), u_n) \tag{3}$$

This constraint implies that the dimensionality of the encoder output should be the same as that of the input. However, if the input dimensionality is smaller than the desired output dimensionality, this could be solved using e.g. random projections, which work well for increasing the dimensionality of neural encoder inputs (Wieting and Kiela, 2019).

Conceptual grounding The *conceptual grounding* objective is a modification of the conceptual meaningfulness objective of the BNE model. The conceptual meaningfulness objective forces the encoding of a biomedical name to be similar to a prototypical representation of its concept. This concept representation is approximated by averaging the pretrained embedding representations of all the names belonging to the concept:

$$u_p = \frac{1}{|C_n|} \sum_{n \in C_n} u_n \tag{4}$$

While converging to this pretrained target is feasible for small synonym sets, such convergence is unnecessary and overfitting for larger classes of names with graded differences in semantic similarity among the class members. To retain the robustness of the encodings, we only want to pull the names in the direction of their pretrained concepts, rather than minimizing their distance entirely. To this end, we simply take the average of the pretrained name representation and the pretrained concept representation:

$$\begin{aligned} v_{ground} &= \frac{u_p + u_n}{2} \\ L_{ground} &= d(f(n), v_{ground}) \end{aligned} \tag{5}$$

Multi-task setup Our multi-task setup sums the losses of the 3 training objectives:

$$L = \alpha L_{sem} + \beta L_{cont} + \gamma L_{ground} \tag{6}$$

where α, β, and γ are possible weights for the individual losses. Since the 3 losses all directly reflect cosine distances, they are similarly scaled and don't require weighting to work properly. In our experiments, $\alpha = \beta = \gamma = 1$ showed the most robust performance along all settings.

4 Data and task setup

4.1 Extracting hierarchical data

Following previous research (Kotitsas et al., 2019; Camacho-Collados et al., 2018), we use IS-A relations between concepts from the SNOMED-CT[1] ontology as biomedical hypo-hypernymy relations. For direct comparison with the publicly released BNE embeddings, which were trained on all disorder concepts of SNOMED-CT, we use the 2018AB release of the UMLS[2] to extract only those SNOMED-CT concepts which are included in the semantic group of disorders[3], and extract their reference terms as disorder names. While the resulting directed graph should be acyclic, there are many inconsistencies, which we resolve by removing all cyclic edges, similar to the naive approach used by Mougin and Bodenreider (2005).

For our experiments, we select 3 different (yet slightly overlapping) subgraphs of IS-A relations by sampling 3 high-level concepts which have around 10K child concepts in our cleaned graph. We extract consistent taxonomies from these subgraphs by removing relations which form shortcuts between otherwise non-consecutive levels of the taxonomy, and by leaving out dead-end concepts which don't have a path to the required level of specification down the taxonomy. Child concepts can have mutually inclusive relations to multiple higher-level concepts on the same level of categorization.

[1] https://www.snomed.org
[2] https://uts.nlm.nih.gov/home.html
[3] https://metamap.nlm.nih.gov/SemanticTypesAndGroups.shtml

C1290864	min	max	mean	stdev
Level 1	1	10203	1015	2053
Level 2	1	10203	291	1101
Level 3	1	3840	118	411
Level 4	1	2607	48	195

Table 2: Descriptive statistics about the number of names per class for the different levels sampled from the subgraph with parent concept C1290864 (*disorder of abdomen*). These statistics show that lower levels have less extreme imbalances between classes.

4.2 Data setup

For each subgraph, we select 4 consecutive levels of parent concepts (level 1 is highest, level 4 is lowest). The concepts on these 4 levels are used as class labels for the names from all concepts below level 4. In other words, names belonging to the parent concepts themselves are not used during training: the parent concepts are only used as reference to cluster the names from the lower levels. Table 1 visualizes an example of this process.

This method of aggregating names can lead to very imbalanced classes. Table 2 shows how large this imbalance can get as we go up the hierarchy. While the training regime of our proposed model should be robust against such data artefacts, we want to take a representative test sample across all classes to empirically validate our approach. Therefore, for multiple iterations, we sample one held-out test name for each class on level 4. This test name is then also used for levels 1-3. Afterwards, we carry out the same procedure to sample validation data for calculating the stopping criterion during training. Table 3 shows the distributions of concepts and names used during training, validation, and testing.

4.3 Task setup

We perform 2 tasks on the held-out SNOMED-CT test data to validate our approach. Evidently, we always evaluate on individual levels of categorization. As intrinsic evaluation, we evaluate trained encoders on semantic similarity ranking. We also include the task of unsupervised hypernym detection as extrinsic evaluation. As we don't use the names of higher-level concepts during training, we can exploit them as previously unobserved hypernymic data to show how much higher-level semantics are being modeled by encoders. If the encoder has learned to represent biomedical semantics more effectively, then the name embedding space can reflect that by being more suited for unsupervised detection of hypernyms.

Table 1 gives examples of hypernym names on all 4 levels. Successful hypernym detection for this data implies e.g. that we rank the previously unobserved hypernym *bite wound* over another previously unobserved hypernym *sting of skin* for the name *tick-borne fever*. This task clearly requires more semantic inference than merely resolving synonyms. In this case, the encoder has to represent that ticks are insects that bite instead of sting.

Semantic similarity ranking We evaluate encoders on the ability to reflect semantic similarity between names by their cosine similarity. Given a mention m of a biomedical name which belongs to the higher-level class c, we have to rank the set of all training names S which includes $C_n \subset S$, a set of training names which belong to the same class c as the test mention. To rank the biomedical names according to their similarity to the mention, we first encode both the mention m as well as every name $n \in S$, and then rank every name n using the cosine similarity between the encoded mention $f(m)$ and the encoded name $f(n)$. We then calculate the Mean Average Precision (mAP) over all test mentions for retrieving training names from the same higher-level class.

Unsupervised hypernym detection Given a test mention m of a biomedical name which belongs to the higher-level class c, we have to rank the set of all hypernym names H belonging to a specific level of categorization. This set includes $C_h \subset H$, the set of hypernym names which belong to the same class c as the test mention. To rank the biomedical names according to their similarity to the mention, we first encode both the mention m as well as every hypernym name $h \in H$, and then rank every hypernym name h using the cosine similarity between the encoded mention $f(m)$ and the encoded hypernym $f(h)$. We then calculate the Mean Reciprocal Rank (MRR) over all test mentions for retrieving hypernym names from the same higher-level class.

5 Experiments and results

5.1 Reference model and baselines

We compare our DAN model against the the publicly released **pretrained BNE** model with skip-gram word embeddings, BNE + SG_w,[4] which was

[4] https://github.com/minhcp/BNE

	C1290864 *disorder of abdomen*	**C0560169** *osteoarthropathy*	**C0263661** *dermatological finding*
Level 1	27	30	35
Level 2	98	86	80
Level 3	248	236	231
Level 4	610	536	602
Lower-level names	24737 / 1557 / 763	20574 / 1335 / 649	25659 / 1567 / 814

Table 3: An overview of the distribution of higher-level classes for the 3 subgraphs used in our experiments. The lower-level names are divided into train / test / validation.

trained on approximately 16K synonym sets of disease concepts in the UMLS, containing 156K disease names. We also include 2 baselines: our 300-dimensional **fastText** name embeddings (defined in Equation 1 in Section 3.1), and averaged 728-dimensional context-specific token activations extracted from the publicly released **BioBERT** model (Lee et al., 2019).

5.2 Training and implementation details

The DAN model is implemented in PyTorch (Paszke et al., 2019). Both the input and output dimensionality are 300 (which is the dimensionality of the input fastText embeddings described in Section 3.1). All encoders for which we report results are finetuned to one hidden layer, which has 76,800 dimensions. Adam optimization (Kingma and Ba, 2015) is performed on a batch size of 64, using a learning rate of 0.001 and a dropout rate of 0.5. Input strings are first tokenized using the Pattern tokenizer (Smedt and Daelemans, 2012) and then lowercased. We use a triplet margin of 0.1 for the siamese triplet loss L_{sem} defined in Equation 2.

To train the model, we iterate over all names in the training data and apply the 3 training objectives for each name in a batch. To avoid overfitting on the largest classes, we always sample one siamese triplet per name, using random sampling for the positive name and distance-weighted sampling for the negative name. As stopping criterion we use the mAP of semantic similarity ranking (as defined in Section 4.3) for held-out validation names: we stop training once this score hasn't improved anymore over 10 epochs. This relaxed stopping criterion allows the model to optimize the subsampled siamese triplet loss in a balanced stochastic way over many epochs without quitting too early.

5.3 Results and discussion

Semantic similarity ranking Table 4 shows the test performance for semantic similarity ranking. First and foremost, the robustness of the Level 1 DAN models is consistently great for all 3 subgraphs. For instance, in the case of the subgraph C1290864 (*disorder of abdomen*), the DAN is trained on only 27 large classes but outperforms the fastText baseline for the 610 classes on Level 4. Secondly, all DAN models generalize both bottom-up and top-down along the hierarchical levels to the extent that they consistently outperform the fastText baseline by a substantial margin.

Thirdly, the slight superiority of BioBERT over fastText for this task is most pronounced for the lowest levels. As we go up in the hierarchy, the difference grows smaller, which leads us to believe that the improvements are not so much of a semantic nature. Interestingly, the pretrained BNE model is competitive with our DAN models for the lower levels, which are still more coarse-grained than the fine-grained distinctions on which the BNE was trained. However, such a bottom-up effect is lacking for the highest levels of categorization. These observations reinforce the notion that both the size (the BNE was trained on 156K disorder names, our models on 20-25K) and the granularity of the data matter for deep representation learning.

Unsupervised hypernym detection Table 5 shows the test performance for unsupervised hypernym detection. These results clearly show trends which are similar to the semantic similarity ranking. Most remarkably, the bottom-up and bottom-down effects are almost as consistent here: the highest-level DAN still outperforms the baselines for the lowest levels and vice versa. One major difference with the results for semantic similarity ranking is the relatively worse performance from BioBERT

	C1290864				C0560169				C0263661			
	1	2	3	4	1	2	3	4	1	2	3	4
DAN level 1	**0.57**	0.50	0.39	0.43	**0.70**	0.44	0.36	0.37	**0.64**	<u>0.55</u>	0.36	0.36
DAN level 2	<u>0.49</u>	**0.58**	0.46	0.48	<u>0.55</u>	**0.58**	0.44	0.44	<u>0.58</u>	**0.59**	0.40	0.39
DAN level 3	0.43	<u>0.51</u>	**0.56**	0.54	0.51	<u>0.51</u>	**0.52**	0.54	0.52	0.52	**0.51**	0.48
DAN level 4	0.38	0.43	<u>0.47</u>	**0.60**	0.45	0.45	<u>0.48</u>	<u>0.58</u>	0.45	0.44	<u>0.41</u>	**0.54**
fastText	0.26	0.27	0.25	0.33	0.36	0.29	0.28	0.32	0.33	0.30	0.24	0.30
BioBERT	0.27	0.29	0.29	0.39	0.38	0.32	0.31	0.37	0.36	0.33	0.27	0.35
BNE	0.35	0.41	0.42	<u>0.57</u>	0.43	0.41	0.45	**0.59**	0.44	0.44	0.39	<u>0.51</u>

Table 4: Test performance of semantic similarity ranking per level, as measured by mAP. The highest score per level of each subgraph is denoted in bold; the second highest score is underlined.

	C1290864				C0560169				C0263661			
	1	2	3	4	1	2	3	4	1	2	3	4
DAN level 1	**0.60**	<u>0.58</u>	0.59	0.68	**0.48**	<u>0.54</u>	0.52	0.63	**0.52**	<u>0.57</u>	0.55	0.62
DAN level 2	0.52	**0.59**	0.62	0.70	<u>0.45</u>	**0.58**	0.56	0.67	<u>0.50</u>	**0.60**	0.57	0.63
DAN level 3	<u>0.55</u>	0.57	**0.66**	<u>0.73</u>	0.41	<u>0.54</u>	**0.58**	<u>0.70</u>	0.48	<u>0.57</u>	**0.62**	0.67
DAN level 4	0.53	0.52	<u>0.63</u>	**0.74**	0.39	0.53	**0.58**	**0.74**	0.46	0.54	<u>0.59</u>	**0.71**
fastText	0.46	0.44	0.53	0.65	0.34	0.47	0.49	0.63	0.38	0.45	0.50	0.59
BioBERT	0.41	0.41	0.50	0.62	0.28	0.41	0.46	0.58	0.39	0.47	0.48	0.59
BNE	0.43	0.50	0.60	0.71	0.42	0.48	0.57	<u>0.70</u>	0.49	0.49	0.54	<u>0.68</u>

Table 5: Test performance for unsupervised hypernym detection per level, as measured by MRR. The highest score per level of each subgraph is denoted in bold; the second highest score is underlined.

here compared to fastText. This is in line with the findings by Yu et al. (2020), who report that BERT does not yield considerable improvement for hypernymy detection in their experiments. It also puts into perspective to what extent we can expect higher-level semantics to be encoded solely through self-supervised methods.

Table 6 gives an example of hypernym rankings for the test mention *poisoning caused by mexican beaded lizard bite*. By clustering similar names together with other bite wounds during training, the DAN model has learned to recognize the test mention as a bite wound. The BNE has failed to do so.

The effectiveness of our unsupervised method using only cosine similarity contrasts with earlier approaches which explicitly require more than cosine similarity to properly work. For example, Vulić and Mrkšić (2018) use vector norms to encode hierarchical hypernymic relations, while other research into hypernymy even requires other geometric spaces than Euclidean space, such as hyperbolic space (Dhingra et al., 2018). Our results can indicate that cosine similarity in Euclidean space still shows potential for encoding these hierarchical relations given the right training objectives.

5.4 Semantic relatedness benchmarks

We also evaluate our name encoders on two biomedical benchmarks of semantic similarity, which allow to compare cosine similarity between name embeddings with human judgments of relatedness. MayoSRS (Pakhomov et al., 2011) contains multi-word name pairs of related but different fine-grained concepts. UMNSRS (Pakhomov et al., 2016) contains only single-word pairs, which also stem from different fine-grained concepts. This benchmark makes a distinction between *similarity* and *relatedness*.

The correlations in Table 7 show that the majority of our trained encoders remain robust out-of-the box, with a large portion of them outperforming the fastText baseline which they use as input. The highest-level model trained on the C0560169 subgraph (*dermatological finding*) is even competitive with the pretrained BNE, having been trained on only 30 classes. All in all, these results confirm that our proposed model is relatively robust against variable granularity of clustering, and is not overly tailored to the data artefacts of one specific sub-

Subgraph	C0560169	
Level	3	
Test mention	**poisoning caused by mexican beaded lizard bite**	
Matching hypernyms	bite wound / bite wound (disorder)	
	DAN Level 1	**BNE**
Top 5 ranking	*bite wound (disorder)*	infestation caused by fly larvae (disorder)
	bite wound	fly larva infestation
	open traumatic dislocation of hip, unspecified	infestation caused by fly larvae
	open traumatic dislocation of hip, unspecified (disorder)	infestation by fly larvae (disorder)
	open dislocation of phalanx of foot (disorder)	infestation by fly larvae

Table 6: A comparison between our DAN encoder and the BNE reference model for unsupervised hypernym ranking of the Level 3 test mention *poisoning caused by mexican beaded lizard bite*. The DAN model generalizes from the training data to associate the test mention correctly with bite wounds. In the training process, it seems to have clustered bite wounds together with open dislocations. The BNE model apparently associates lizards with infestations by fly larvae, but fails to recognize that there is a bite wound mentioned in the test mention.

graph.

5.5 Discussion

While our empirical results are certainly encouraging, the true robustness of our proposed framework remains an open question. Whereas our proposed DAN model remains robust over entire hierarchies for semantic similarity ranking and unsupervised hypernym detection, its relative performance for the semantic relatedness benchmarks is not entirely predictable from those tasks. One the one hand, this likely has to do with the modest sizes of the benchmarks, for which small to very small margins in performance are not very reliable or indicative.

On the other hand, we also have to consider that our finetuned DAN only contains a single, yet very wide, hidden layer. This implies that the encoder network relies more on what can considered to be an elaborate weighted average than a deep multilayer transformation of the input. While this is not very surprising in the context of transferable representations (and emphasizes the effectiveness of exploiting word embeddings according to their full potential in simple ways, as suggested by Wieting and Kiela (2019)), it still raises the question whether there are straightforward regularization alternatives to the contextual meaningfulness objective which can allow for deep transformations with the DAN.

	MayoSRS (rel)	UMNSRS (rel)	UMNSRS (sim)
fastText	0.44	0.47	0.48
Level 1 C0560169	0.42	<u>0.55</u>	<u>0.54</u>
Level 2 C0560169	0.47	0.51	0.50
Level 3 C0560169	0.50	0.51	0.50
Level 4 C0560169	0.50	0.51	0.50
Level 1 C1290864	0.52	0.42	0.46
Level 2 C1290864	0.55	0.46	0.40
Level 3 C1290864	0.53	0.46	0.50
Level 4 C1290864	<u>0.56</u>	0.45	0.50
Level 1 C0263661	0.46	0.49	0.51
Level 1 C0263661	0.51	0.47	0.50
Level 3 C0263661	0.55	0.50	0.53
Level 4 C0263661	0.52	0.50	0.50
Phan et al. (2019)	**0.63**	**0.58**	**0.61**

Table 7: Spearman's rank correlation coefficient between cosine similarity scores of name embeddings and human judgments, reported on semantic similarity (sim) and relatedness (rel) benchmarks. The highest score is denoted in bold; the second highest is underlined.

6 Conclusion and future work

In this paper, we have introduced the challenge of integrating higher-level semantics into robust biomedical name representations. We provide a framework to both train and evaluate encoders for this task. Moreover, we have proposed a modification of the Biomedical Name Encoder model which is directly applicable to a variety of coarse-grained categorizations. This modification replaces more complex neural architectures with a lightweight Deep Averaging Network encoder, which is easily scalable to the large amounts of required training data, while remaining sufficiently robust. The only important hyperparameter to tune for this encoder is the size of the Feedforward Neural Network.

Experiments indicate that our proposed framework can even be effective using only around 30 coarse-grained classes. This opens up possibilities for applying our framework to data beyond carefully curated ontologies, for instance in self-supervised or semi-supervised settings. Future work will try to understand and define the limits of applying our framework to such settings.

Acknowledgments

We would like to thank the anonymous reviewers for their feedback. This research was funded by the government agency Flanders Innovation & Entrepreneurship (VLAIO) in the framework of the Accumulate VLAIO SBO project, and by the Flemish Government (AI Research Program).

References

Piotr Bojanowski, Edouard Grave, Armand Joulin, and Tomas Mikolov. 2017. Enriching word vectors with subword information. *Transactions of the Association for Computational Linguistics*, 5:135–146.

Jose Camacho-Collados, Claudio Delli Bovi, Luis Espinosa-Anke, Sergio Oramas, Tommaso Pasini, Enrico Santus, Vered Shwartz, Roberto Navigli, and Horacio Saggion. 2018. SemEval-2018 task 9: Hypernym discovery. In *Proceedings of The 12th International Workshop on Semantic Evaluation*, pages 712–724, New Orleans, Louisiana. Association for Computational Linguistics.

Gal Chechik, Varun Sharma, Uri Shalit, and Samy Bengio. 2010. Large scale online learning of image similarity through ranking. *Journal of Machine Learning Research*, 11:1109–1135.

Bhuwan Dhingra, Christopher Shallue, Mohammad Norouzi, Andrew Dai, and George Dahl. 2018. Embedding text in hyperbolic spaces. In *Proceedings of the Twelfth Workshop on Graph-Based Methods for Natural Language Processing (TextGraphs-12)*, pages 59–69, New Orleans, Louisiana, USA. Association for Computational Linguistics.

Alex Graves and Jurgen Schmidhuber. 2005. Framewise phoneme classification with bidirectional LSTM and other neural network architectures. *Neural Networks*, 18(5-6):602–610.

Yanming Guo, Yu Liu, Erwin M. Bakker, Yuanhao Guo, and Michael S. Lew. 2017. CNN-RNN: a large-scale hierarchical image classification framework. *Multimedia Tools and Applications*, 77:10251–10271.

Kai Hakala, Suwisa Kaewphan, Tapio Salakoski, and Filip Ginter. 2016. Syntactic analyses and named entity recognition for PubMed and PubMed Central — up-to-the-minute. *Proceedings of the 15th Workshop on Biomedical Natural Language Processing*, pages 102–107.

Peter Hase, Chaofan Chen, Oscar Li, and Cynthia Rudin. 2019. Interpretable image recognition with hierarchical prototypes. In *The Seventh AAAI Conference on Human Computation and Crowdsourcing (HCOMP-19)*.

Mohit Iyyer, Varun Manjunatha, Jordan Boyd-Graber, and Hal Daumé III. 2015. Deep unordered composition rivals syntactic methods for text classification. In *Association for Computational Linguistics*.

Katikapalli Subramanyam Kalyan and Sivanesan Sangeetha. 2020. Medical concept normalization in user-generated texts by learning target concept embeddings. In *Proceedings of the 11th International Workshop on Health Text Mining and Information Analysis*, pages 18–23, Online. Association for Computational Linguistics.

Dimitri Kartsaklis, Mohammad Taher Pilehvar, and Nigel Collier. 2018. Mapping text to knowledge graph entities using multi-sense LSTMs. *Proceedings of the 2018 Conference on Empirical Methods in Natural Language Processing*, pages 1959–1970.

Diederik P. Kingma and Jimmy Ba. 2015. Adam: A method for stochastic optimization. In *3rd International Conference for Learning Representations (ICLR)*.

Sotiris Kotitsas, Dimitris Pappas, Ion Androutsopoulos, Ryan McDonald, and Marianna Apidianaki. 2019. Embedding biomedical ontologies by jointly encoding network structure and textual node descriptors. In *Proceedings of the 18th BioNLP Workshop and Shared Task*, pages 298–308, Florence, Italy. Association for Computational Linguistics.

P. H. Le-Khac, G. Healy, and A. F. Smeaton. 2020. Contrastive representation learning: A framework and review. *IEEE Access*, 8:193907–193934.

Jinhyuk Lee, Wonjin Yoon, Sungdong Kim, Donghyeon Kim, Sunkyu Kim, Chan Ho So, and Jaewoo Kang. 2019. BioBERT: a pre-trained biomedical language representation model for biomedical text mining. *Bioinformatics*, 36(4):1234–1240.

Fleur Mougin and Olivier Bodenreider. 2005. Approaches to eliminating cycles in the UMLS metathesaurus: Naïve vs. formal. In *AMIA Annual Symposium Proceedings*.

Serguei V.S. Pakhomov, Greg Finley, Reed McEwan, Yan Wang, and Genevieve B. Melton. 2016. Corpus domain effects on distributional semantic modeling of medical terms. *Bioinformatics*, 32(23):3635–3644.

Serguei V.S. Pakhomov, Ted Pedersen, Bridget McInnes, Genevieve B. Melton, Alexander Ruggieri, and Christopher G. Chute. 2011. Towards a framework for developing semantic relatedness reference standards. *Journal of Biomedical Informatics*, 44:251–265.

Adam Paszke, Sam Gross, Francisco Massa, Adam Lerer, James Bradbury, Gregory Chanan, Trevor Killeen, Zeming Lin, Natalia Gimelshein, Luca Antiga, Alban Desmaison, Andreas Kopf, Edward Yang, Zachary DeVito, Martin Raison, Alykhan Tejani, Sasank Chilamkurthy, Benoit Steiner, Lu Fang, Junjie Bai, and Soumith Chintala. 2019. PyTorch: An imperative style, high-performance deep learning library. In H. Wallach, H. Larochelle, A. Beygelzimer, F. d'Alché-Buc, E. Fox, and R. Garnett, editors, *Advances in Neural Information Processing Systems 32*, pages 8024–8035. Curran Associates, Inc.

Minh C. Phan, Aixin Sun, and Yi Tay. 2019. Robust representation learning of biomedical names. In *Proceedings of the 57th Annual Meeting of the Association for Computational Linguistics*, pages 3275–3285, Florence, Italy. Association for Computational Linguistics.

Dinghan Shen, Guoyin Wang, Wenlin Wang, Martin Renqiang Min, Qinliang Su, Yizhe Zhang, Chunyuan Li, Ricardo Henao, and Lawrence Carin. 2018. Baseline needs more love: On simple word-embedding based models and associated pooling mechanisms. In *Proceedings of the 56th Annual Meeting of the Association for Computational Linguistics (Long Papers)*, pages 440–450.

Tom De Smedt and Walter Daelemans. 2012. Pattern for Python. *Journal of Machine Learning Research*, 13:2031–2035.

Mujeen Sung, Hwisang Jeon, Jinhyuk Lee, and Jaewoo Kang. 2020. Biomedical entity representations with synonym marginalization. In *Proceedings of the 58th Annual Meeting of the Association for Computational Linguistics*, pages 3641–3650, Online. Association for Computational Linguistics.

Fariborz Taherkhani, Hadi Kazemi, Ali Dabouei, Jeremy E. Dawson, and Nasser M. Nasrabadi. 2019. A weakly supervised fine label classifier enhanced by coarse supervision. In *ICCV*.

Noha S. Tawfik and Marco R. Spruit. 2020. Evaluating sentence representations for biomedical text: Methods and experimental results. *Journal of Biomedical Informatics*, 104.

Ivan Vulić and Nikola Mrkšić. 2018. Specialising word vectors for lexical entailment. In *Proceedings of the 2018 Conference of the North American Chapter of the Association for Computational Linguistics: Human Language Technologies, Volume 1 (Long Papers)*, pages 1134–1145, New Orleans, Louisiana. Association for Computational Linguistics.

John Wieting and Douwe Kiela. 2019. No training required: Exploring random encoders for sentence classification. In *International Conference on Learning Representations*.

Chao-Yuan Wu, R. Manmatha, Alexander J. Smola, and Philipp Krahenbuhl. 2017. Sampling matters in deep embedding learning. In *ICCV*.

Cinna Wu, Mark Tygert, and Yann LeCun. 2019. A hierarchical loss and its problems when classifying non-hierarchically. *PLOS ONE*, 14(12):1–17.

Changlong Yu, Jialong Han, Peifeng Wang, Yangqiu Song, Hongming Zhang, Wilfred Ng, and Shuming Shi. 2020. When hearst is not enough: Improving hypernymy detection from corpus with distributional models. In *Proceedings of the 2020 Conference on Empirical Methods in Natural Language Processing (EMNLP)*, pages 6208–6217, Online. Association for Computational Linguistics.

Classification of mental illnesses on social media using RoBERTa

Ankit Murarka*
IBM / Raleigh, NC
ankit.murarka1@ibm.com

Balaji Radhakrishnan*
balag59@gmail.com

Sushma Ravichandran*
IBM Research / Yorktown heights, NY
sushma.ravichandran@ibm.com

Abstract

Given the current social distancing regulations across the world, social media has become the primary mode of communication for most people. This has isolated millions suffering from mental illnesses who are unable to receive assistance in person. They have increasingly turned to online platforms to express themselves and to look for guidance in dealing with their illnesses. Keeping this in mind, we propose a solution to classify mental illness posts on social media thereby enabling users to seek appropriate help. In this work, we classify five prominent kinds of mental illnesses- depression, anxiety, bipolar disorder, ADHD and PTSD by analyzing unstructured user data on Reddit. In addition, we share a new high-quality dataset[1] to drive research on this topic. The dataset consists of the title and post texts from 17159 posts and 13 subreddits- each associated with one of the five mental illnesses listed above or a `None` class indicating the absence of any mental illness. Our model is trained on Reddit data but is easily extensible to other social media platforms as well as demonstrated in our results.We believe that our work is the first multi-class model that uses a Transformer Vaswani et al. (2017)-based architecture such as RoBERTa Liu et al. (2019) to analyze people's emotions and psychology. We also demonstrate how we stress-test our model using behavioral testing. Our dataset is publicly available and we encourage researchers to utilize this to advance research in this arena. We hope that this work contributes to the public health system by automating some of the detection process and alerting relevant authorities about users that need immediate help.

* These authors contributed equally

[1]https://github.com/amurark/mental-health-detection

1 Introduction

During these unprecedented times when the world is plagued by COVID19, a large number of people have been showing symptoms of clinical anxiety or depression[2]. This can be attributed to a myriad of reasons including lock down, mandatory social distancing, higher unemployment, economic depression and work-related stress.

In a report published earlier this year, the American Foundation for Suicide Prevention found that people experience anxiety (53%) and sadness (51%) more often now than before the coronavirus pandemic.

In the past decade, social media has transformed how people interact with each other. Apart from sharing factual information and news, people actively partake in sharing their day to day activities, experiences, feelings, opinions, hopes, desires, and emotions online. These texts provide information which can be used to identify the mental health individuals. Furthermore, the current state of enforced social distancing and isolation has propelled more people to express their emotions on social media as it provides them with an accessible platform to share their thoughts with others, many a times, in search for help.

Our research work utilizes user data, especially the kind pertaining to emotions as this class of data can give us valuable insights about the mental state of a person; and, in turn, our work has the potential to assist in the diagnosis and analysis of various mental disorders. This study aims to bridge the gap between people in search of help and experts who can provide the needed help.

Due to the paucity of adequate annotated and structured user data in this domain, we decided to generate our own dataset by crawling subreddits

[2]https://afsp.org

Proceedings of the 12th International Workshop on Health Text Mining and Information Analysis, pages 59–68
April 19, 2021.

on reddit.com[3] pertaining to our use case as a lot of users were found to have shared their feelings there. Although Reddit was the sole source of our dataset, we believe that this study can be seamlessly extended to other social media platforms as well because of the presence of similar unstructured user data online.

The advent of transformers and BERT Devlin et al. (2018) has caused quite a stir in the NLP community because of the state-of-the-art results it was able to produce in various NLP tasks. In this work, we use a RoBERTa Liu et al. (2019) based classifier, which has a similar architecture to BERT with an improved pre-training procedure. RoBERTa's effective and efficient performance on unstructured data and its ability to learn contextual information compelled us to explore and capitalize its power to categorize online user generated texts into various classes of Mental Illness.

We identify five broad classes of mental illnesses - depression, anxiety, bipolar disorder, ADHD (Attention Deficit Hyperactivity Disorder), PTSD (Post Traumatic Stress Disorder) and an additional 'None' class (which does not pertain to any mental illness). We train a multi-class classifier on the data crawled from online user data. Based on our experiments, we present encouraging results that demonstrate that social media data has the potential to complement standard clinical procedures in the prognosis of mental health amongst two broad categories of users - ones who are seeking help online and ones who are unbeknownst of their condition.

2 Related Work

In the recent past, people have increasingly turned to social media to share and seek counsel on the topic of mental health. This has prompted researchers to utilize the information and apply a plethora of techniques in NLP and Machine Learning in order to assist people who might require help. Most of the recent such research, as described in the subsequent paragraphs, has revolved around Reddit data: Kim et al. (2020), Gkotsis et al. (2017), Sekulic and Strube (2019), Zirikly et al. (2019). Prior to this recent shift to Reddit data, a lot of the earlier research was focused on utilizing Twitter data: Orabi et al. (2018), Benton et al. (2017), Coppersmith et al. (2015).

There have been a wide variety of approaches ranging from classical NLP techniques to neural network based deep learning methods. Coppersmith et al. (2015) used character level language models to examine how likely a sequence of characters is to be generated by a user with mental health issues. Benton et al. (2017) evaluated a standard regression model, a multilayer perceptron single-task learning (STL) model, and a neural MTL model on detecting multiple types of mental health issues. Orabi et al. (2018) utilized word embeddings in tandem with a variety of neural network models like CNNs and RNNs to detect depression. Gkotsis et al. (2017) experimented with Feed Forward Neural Networks, CNNs, SVMs and Linear classifiers to perform binary classification on mental health posts. Sekulic and Strube (2019) came up with the approach of using Hierarchical Attention Networks(HANs) to detect a wide range of mental health issues like Depression, ADHD, Anxiety etc. and trained a binary classifier for each of the disorders. The most recent work on this was by Kim et al. (2020) who proposed a CNN-based classification model. Once again though, each disorder had its own separate binary classifier to perform the detection.

To our knowledge, this is the first attempt at treating this problem as a multi-class classification problem, where a single classifier can accurately classify the type of disorder that the person is referring to in their post. In addition, this is also the first work that harnesses the incredible capabilities of an advanced Transformer based algorithms like RoBERTa to solve this difficult problem.

3 Dataset

The Reddit API was used to crawl 13 Reddit Subreddits for a total of 17159 posts (text and title) to obtain the data for this work. The text from comment threads of these posts was not collected as it tended to diverge from the main topic of the subreddit. Even though there are a lot of mental illnesses that need addressing, only 5 of them had sufficient data for our purposes that were chosen for the purposes of this paper. They are: `bipolar`, `adhd`, `anxiety`, `depression` and `ptsd`. We do plan to extend this solution to tackle the remaining illnesses as well in due time. The posts in these subreddits were assigned a class label corresponding to the name of the mental illness they were associated with. All the remaining subreddits were carefully chosen as to minimize any chances of thematic content overlap between them and the illness classes

[3]https://www.reddit.com

i.e to enable the model to differentiate between posts discussing mental illness from those that were not, we chose a few subreddits from a wide range of common topics such as `music`, `travel`, `india`, `politics`, `english`, `datasets`, `mathematics` and `science`. The text from these subreddits were combined and assigned the class label None. While we are currently treating this task as a multi-class classification problem, we duly acknowledge the fact that mental illnesses are generally co-related and require multi-label classification techniques. We are actively working on converting this task to a multi-label classification problem.

While collecting data, we ensured that the number of upvotes for each post in all subreddits is more than 10. We also set a minimum post token length of 30 tokens. These numbers were chosen after carefully perusing through the data and going over the post text in order to retain quality in the dataset. We initially crawled the subreddits pertaining to mental illness and were able to collect about 3000 posts in total. We then crawled the other subreddits corresponding to the None class label and collected about 300 posts per subreddit for 8 subreddits. This also ensured a good balance of class labels. While selecting the eight general topic subreddits, we not only selected subreddits that have sufficiently high number of posts, but also ensured that we cover a broad range of topics. Table 1 shows the statistics collected for each subreddit. The dataset was preprocessed to remove any URLs or usernames that could potentially contain sensitive information. This was done keeping in mind that the dataset will be released publicly for the purpose of extending this research work.

To gauge the data quality we ran some analysis. We manually went over the lowest voted posts for each mental illness subreddit. We wanted to establish that texts from these posts expressed emotions from people discussing corresponding mental illness it is labelled as. Table 2 presents excerpts of lowest voted post from each mental illness subreddit.

We also certified that the general topic subreddits did not have a high similarity with the posts corresponding to other the other 5 subreddits. This was done to ensure that we do not have any false negatives while assigning truth labels. We counted the number of posts the mental illness terms appeared in, for each subreddit. The subreddits corresponding to mental illnesses had a much higher count of these words. In addition to this, we compared the cosine similarity between some of the highest/lowest posts of mental illness subreddits and the general topic subreddits and manually compared the results to find that this distance was higher than the distance between two posts of the mental illness subreddits.

We also attempted to augment the data using Easy Data Augmentation Wei and Zou (2019) to boost the performance of our model. However, we did not observe an apparent shift in our evaluation metrics- explained by the fact the EDA is meant to perform best for smaller datasets ($\leq$ 5000 sample sizes).

4 Model

In this section, we describe our model architecture for the multi-class mental illness classification task. We propose a RoBERTa based classifier in order to accomplish this. In addition, we also compare the proposed model against an LSTM Hochreiter and Schmidhuber (1997) based classifier and a BERT Devlin et al. (2018) based classifier to demonstrate the superiority of our approach. Since this is an entirely new dataset, there is no established baseline, so the LSTM model will serve as the baseline for our experiments. We also showcase our gains over BERT, the most widely used transformer model today for text classification. All our models were implented in Pytorch Paszke et al. (2019). The Transformer models were implemented with the help of the HuggingFace Transformers Wolf et al. (2019) library.

4.1 LSTM based classifier

LSTMs(Long Short-Term Memory) were the state of the art models when it came to text classification before the advent of Transformers. They will serve as our baseline. First we tokenized the sentences using NLTK[4] and converted them to lower case to create our vocabulary. In order to get rid of words that might not exist, we removed all words from our vocabulary that appear only once. We also added `padding` and `unknown` to our vocabulary in order to account for padding and unknown tokens respectively. Each sentence was represented using a sequence of length 512 and this forms our input to the LSTM model. We used a 2 layer LSTM for all our experiments with an embedding layer of size 100 and a hidden layer size of 256. Dropout

[4] https://www.nltk.org/

Subreddit	Number of posts	Average no. of words (posts)	Average no. of words (titles)	Average Upvotes	Highest Upvotes	Lowest Upvotes
r/depression	3062	152.74	12.20	517.19	4802	11
r/anxiety	3027	170.38	11.75	246.07	3349	11
r/ptsd	2501	233.55	10.14	38.4	443	11
r/adhd	3082	198.55	13.71	377.13	4484	11
r/bipolar	3009	203.28	9.26	32.37	363	11
none	2478	238.52	15.76	6715.33	199295	11

Table 1: Dataset: Statistics

Lowest rated post
r/depression- The older I am getting the less hope I have to secure a life worth living. I feel finished because I always had that state of mind **r/adhd**- Does anyone else feel like u do have a personality and ability to make friends but ure kind of stuck in ur own body **r/bipolar**- I just made a really impulsive choice with my breed of dog because I had to have one NOW. I was thinking about it constantly day and night and I couldn't sleep. **r/anxiety**- I find myself constantly remembering embarrassing or cringey moments from my past (ranging anywhere from present day to back about 10 years) and cringing hard at them **r/ptsd**- I feel like I am just constantly angry. Angry about my trauma and how it has affected me, and angry about where I am in my life because of it. I don't want to be angry anymore

Table 2: Dataset: Posts excerpts

Srivastava et al. (2014) with a probability of 0.5 was used in order to achieve regularization. We used standard cross-entropy loss as the loss function. During training, Adam Kingma and Ba (2014) was the optimizer of choice, with a learning rate of 0.005. The model was trained for a total of 25 epochs with a batch size of 32. Gradient clipping was used to prevent exploding gradients.

4.2 BERT based classifier

BERT(Bidirectional Encoder Representations from Transformers) has been the biggest breakthrough in NLP in the recent past with state of the art results in a myriad of tasks. Since its inception, better models with gains have been trickling along, but BERT continues to be the most popular model for text classification even today. The BERT classifier comprises a fine-tuned BERT model followed by a dropout layer and a fully connected layer. We fine-tuned a pre-trained BERT-base model on our dataset for this task. A pre-trained tokenizer on BERT is used to tokenize our input sentences. After carefully examining the sentence length distribution, we chose a sequence length of 35 for titles, and 512 for posts and posts+titles. Either padding or truncation was used to ensure that all sentences were represented using the same sequence length. All the BERT based models were fine-tuned on our data for 10 epochs with a learning rate of 1e-5. Adam served as the optimizer and cross-entropy loss was the loss function of choice. A dropout layer with probability of 0.3 was used for the sake of regularization.

4.3 RoBERTa based classifier

RoBERTa(Robustly Optimized BERT Pretraining Approach) is another state of the art language model that builds on BERT by modifying key hyperparameters and training on more data. It outperforms BERT on several benchmark tasks and forms the core of our proposed solution. In order to make it a fair comparison with BERT, we retain the architecture and all design choices made with the BERT based classifier barring the pretrained model and the tokenizer which are now all based on RoBERTa. The input sentences were tokenized using a pre-trained tokenizer on RoBERTa-base. Just as in the case of BERT, we chose a sequence length of 35 for titles, and 512 for posts and posts+titles. Similar to BERT, the RoBERTa based models were also fine-tuned for 10 epochs with a learning rate of 1e-5 and Adam. A batch size of 32 was used while fine-tuning on the titles whereas a batch size of 16 was the only viable option to fine-tune on posts and posts+titles. Cross-entropy remained the preferred loss function. Once again, a dropout layer with probability of 0.3 was used for regularization.

5 Result Analysis

As described earlier, in addition to our primary RoBERTa classifier, we also run experiments on an LSTM classifier and a BERT classifier for the sake of comparison. We fine-tune each of the aforementioned models on just the titles, just the posts and a combination of both in order to perform comprehensive tests and comparisons. When combining the titles and posts for our Transformer models, we convert the problem into a sequence-pair classification task. This allows the model to give more importance to the title which would otherwise be lost when combining the title and the post into one single input given the relative difference in their lengths (the average number of tokens in titles is roughly 6% that of posts).

The results from our experiments are documented in Tables 3 through 5.

As can be observed from Table 3, our proposed RoBERTa based classifier far outperforms the baseline LSTM in all categories. The BERT classifier has results which are quite close to that of RoBERTa's and both beat LSTM by a significant margin, showcasing the incredible capabilities of pre-trained Transformer based architectures. In fact, our RoBERTa model fine-tuned on just the titles was able match the performance of the LSTM model trained on posts. The RoBERTa model was able to achieve an F1 score of 0.86 on the posts and 0.89 on posts+titles which are extremely promising given the complex nature of the multi-class mental illness classification task. The jump in accuracy between posts and posts+titles is not as drastic as the jump between titles and posts. This indicates that the posts offer far more valuable information when compared to the titles and also the fact that most of the useful and relevant information can be extracted from the posts alone. This strong performance on just the posts bodes well for the extensibility of our approach as this can be applied on almost any given social media post without the need for structure in the data like titles, user names, user history, etc.

The rest of this section will focus solely on the results of our best performing RoBERTa model. Table 4 showcases the granular class-wise results of the RoBERTa model. This table in conjunction with the confusion matrices from Figure 1 offers us a wealth of useful and interpretable information.

The first strikingly obvious result is the high accuracy with which the model is able to detect non-illness related posts. Even with just the titles, the model is able to classify the `none` class with an f1 score of more than 0.9. This gives us hope that this model will suffer from very few false positives when it comes to mental illness detection on social media.

An even more crucial property of our model can be noticed in the confusion matrices for posts and posts+titles in Figure 1. When using posts, just 3 illness related posts across the entire test dataset were misclassified as non-illness posts. This number further reduces to 0 when using titles+posts. This shows that the model will detect mental illness posts correctly nearly every single time, thus ensuring that posts from people who are seeking help never go unnoticed when this solution is deployed in the real world.

When it comes to the class wise performance amongst the mental illnesses, the two best performing classes are `adhd` and `ptsd` whereas the two worst performing classes are `depression` and `anxiety`.

The performance of `depression` and `anxiety` classes can be attributed to a few factors. The average number of words per post for `depression` and `anxiety` are the least for any given class. For instance, `depression` posts have roughly 53% lesser textual data when compared to `ptsd` posts. In addition, studies show that depression might often occur in tandem with another mental illness and our data and results back this up as well. The `depression` word occurs in 12% of `anxiety` posts, 12% of `ptsd` posts and 31% of `bipolar` posts. Similarly, `anxiety` occurs in 20% of `ptsd` posts, 12% of `adhd` posts and 14% of `bipolar` posts. This implies that the model cannot give high importance to the mention of these class names like it can with rest of the illnesses, thus making the classification of these 2 classes that much harder.

This can also explain the relatively lower precision scores(higher number of False Positives) for `depression` and `anxiety`. When the other illnesses(excluding `depression` and `anxiety`) are misclassified, they are almost always misclassified as either `depression` or `anxiety`, as can be viewed

[5] In all our tables, P and R stand for Precision and Recall respectively

Models	posts				titles				posts+titles			
	P	R	F1	Acc	P	R	F1	Acc	P	R	F1	Acc
LSTM	0.74	0.72	0.72	0.72	0.65	0.64	0.64	0.64	0.77	0.76	0.76	0.76
BERT	0.83	0.82	0.82	0.82	0.72	0.71	0.71	0.71	0.87	0.87	0.87	0.87
RoBERTa	0.86	0.86	**0.86**	0.86	0.73	0.72	**0.72**	0.72	0.89	0.89	**0.89**	0.89

Table 3: Results: Classification Report

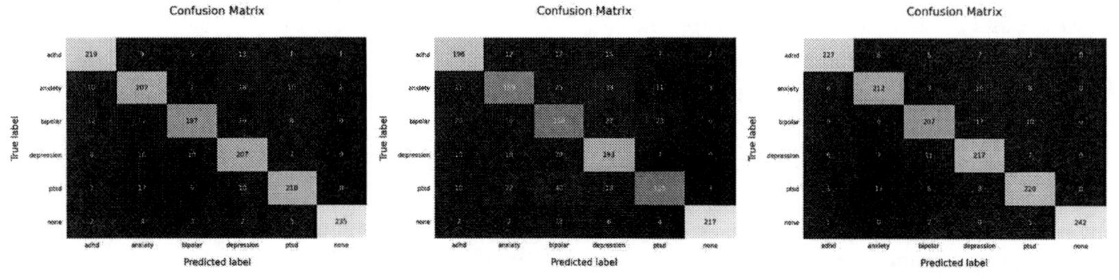

(a) Input: posts (b) Input: titles (c) Input: posts+titles

Figure 1: RoBERTa: Confusion Matrices

Class	posts			titles			posts+titles		
	P	R	F1	P	R	F1	P	R	F1
adhd	0.87	0.88	0.87	0.77	0.79	0.78	0.91	0.92	0.91
anxiety	0.78	0.83	0.81	0.69	0.64	0.67	0.87	0.85	0.86
bipolar	0.88	0.79	0.83	0.58	0.63	0.60	0.88	0.83	0.86
depression	0.77	0.83	0.80	0.65	0.78	0.71	0.81	0.88	0.84
ptsd	0.88	0.85	0.86	0.75	0.62	0.68	0.88	0.89	0.88
none	0.99	0.95	0.97	0.94	0.88	0.91	1.00	0.98	0.99

Table 4: Results: RoBERTa Class-wise results

in Figure 1. In the same figure, we can see that `depression` and `anxiety` are often misclassified as each other due to the reason that they commonly occur together.

There are more posts in the `adhd` and `ptsd` classes that mention the words `depression` and `anxiety` than their respective class names itself. One would assume that this would result in subpar results, but, these classes actually perform the best. This really showcases the true potential of our model, where it doesn't just rely on mention of class names, but has a strong understanding of the context of the post itself. Additionally, the symptoms or descriptions provided for these classes could be strong, unique and discriminative enough for the model to be able to classify them correctly even with all the mentions of other class names.

In Table 5 we have documented a few interesting results we observed in the test set. In the first two examples on the table, the RoBERTa model was able to classify the posts correctly without the presence of class names in the input. The prediction is based purely on contextual information learnt about the class labels during the training process. The next two results are interesting because the truth label assigned to the input text may or may not correspond to actual mental illness described in the text. Since, we are not domain experts ourselves, we would need expert intervention to substantiate this theory. As a part of future work, getting professionals to annotate our dataset might help strengthen the model for such examples. Further samples from our qualitative testing on various social media platforms can be found on our accompanying web-page[6].

Input	Actual	Predicted
often times i'll get distracted from my thoughts either by external influences or just another idea coming in, and then i have to spend a good 5 minutes trying to work out what i was thinking about again.	adhd	adhd
once i come down from flashbacks or panic attacks, i get really bad disassociation. sometimes lasting for days. does anyone else go through this. any tips on how to stop it. i tried grounding but im so far gone it doesn't help.	ptsd	ptsd
i can't sit still when i get my eyebrows done, and when i'm in class i usually doodle to focus. i pay attention very well in school regardless of that, and drawing helps me focus.	anxiety	adhd
i'm flying from dallas to hong kong in january and it's 17 hours. i've flown 12-13 hour flights before and they really mess with me. so i'm wondering - what are your tips for not going crazy on such a long flight? ps: i'm terrible at sleeping on planes. thinking about taking some sleepy meds to see if it'll help	none	anxiety

Table 5: Results: Interesting Examples

Synonym Replacement					**Label Removal**				
Test Set Modified	**posts**				**Test Set Modified**	**posts**			
	P	R	F1	Acc		P	R	F1	Acc
10%	0.86	0.85	0.85	0.85	10%	0.85	0.84	0.84	0.84
50%	0.85	0.84	0.84	0.84	50%	0.81	0.80	0.80	0.80
100%	0.83	0.83	0.83	0.83	100%	0.75	0.74	0.75	0.74
Test Set Modified	**titles**				**Test Set Modified**	**titles**			
	P	R	F1	Acc		P	R	F1	Acc
10%	0.73	0.72	0.72	0.72	10%	0.72	0.71	0.71	0.71
50%	0.71	0.71	0.71	0.71	50%	0.67	0.67	0.67	0.67
100%	0.68	0.67	0.67	0.67	100%	0.61	0.61	0.60	0.61
Label Replace: 'illness'					**Label Replace: random**				
Test Set Modified	**posts**				**Test Set Modified**	**posts**			
	P	R	F1	Acc		P	R	F1	Acc
10%	0.84	0.83	0.84	0.83	10%	0.83	0.82	0.83	0.8
50%	0.78	0.77	0.77	0.77	50%	0.71	0.71	0.71	0.71
100%	0.70	0.67	0.68	0.67	100%	0.58	0.57	0.57	0.57
Test Set Modified	**titles**				**Test Set Modified**	**titles**			
	P	R	F1	Acc		P	R	F1	Acc
10%	0.72	0.71	0.71	0.71	10%	0.71	0.71	0.71	0.71
50%	0.67	0.65	0.65	0.65	50%	0.64	0.64	0.64	0.64
100%	0.62	0.57	0.58	0.57	100%	0.54	0.54	0.54	0.54

Table 6: Behavioral Tests

6 Behavioral Testing

Although the classification metrics analyzed in the previous section are generally regarded sufficient in estimating the performance of Bert-based models, a recent inclination of NLP researchers to behavioral testing inspired us to stress test our models as well.

For all our tests, we used our proposed RoBERTa model and applied these tests to inputs that were either titles or posts. Since we hope to extend our model to other social media platforms, we do not always expect input texts to have a title as well as a descriptive text/post. We adopted the Checklist approach Ribeiro et al. (2020) which involve tests conducted to comprehensively analyze the model's performance.

6.1 Synonym Replacement

Synonym replacement is a kind of Invariance Test where label-preserving perturbations are made to the test set. As labels, the root form of the mental illnesses was chosen- `depress`, `ptsd`, `anxiou/anxiet`, `bipolar` and `adhd`. Python's NLTK package and WordNet were used for these tests.

This test is conducted such that the root words are not perturbed when modifying the test set. For each post, 10% of the tokens were randomly selected (not including the stop words or the root words). Each token was then replaced with one of its synonyms. We used the same logic for titles. We set a max and min on the number of tokens to be selected for replacement - this was (4, 30) for posts and (1, 5) for titles. Since each token was replaced with a synonym, the class label for the samples was not changed. We did this for 10, 50 and 100 percent of the test set and observed results.

In all three cases, we expect the classification metrics to drop. For the case when 10% of the test case was modified the drop was much lower as compared to when 100% of the test case was modified. The results are documented in Table 6. When comparing these results to those in Table 4 we find that the drop in each category is about 2-4% for posts and 5-7% for titles. The lower drop can be attributed to the fact that synonym replacement does not alter the semantics of the input text. Therefore the model was able to draw sufficient information from the input.

[6]https://mental-health-classification.github.io/

6.2 Masking

We also performed a Directional Expectation test on the model. This is similar to the previous test but is instead performed only on labels. The labels, as defined in the previous subsection, are a list of the root form of mental illness class labels. We noticed that the root words appear often in our input texts. This behavioral test was performed to observe our model's dependency on these words. For all the tests below, we modified only those tokens that contained a root word.

In the first case, for every post from a subreddit related to a mental illness, the root form of its class label was removed from the input. For example, the input text: *I feel happy for some time and then depressed again. I'm definitely bipolar* from the `r/bipolar` subreddit, was modified to *I feel happy for some time and then depressed again. I'm definitely*. Note that changes were not made to the word *depressed* in the input. The class label for each modified sample was not changed after the perturbations. Like the previous subsection, these tests were performed on 10, 50 and 100% of the test set.

In the second case, instead of entirely removing the tokens, we replaced it with a generic token `illness`. We expected this modification to retain some semantic information that was lost in the previous test. However, we found that adding a generic token introduced some noise which reduced the overall performance of the model.

Lastly, the tokens were replaced by a randomly chosen root form of a mental illness other than its class label. With this test, we expect to force the model to pick between the label and non label tokens during classification. We believe that this is an interesting scenario to observe.

In all three cases (Table 6), the model performance drops by some degree when compared to Table 4. The first two cases showed a somewhat similar performance drop. However, the model performance was worse than that of the Synonym Replacement test. This means that the model depends on the existence of the root words in the input text to some degree.

In the third scenario, we note that the performance drop is higher. Although the test is meant to confuse the model, we observed that in some cases (especially for input: posts), we got an F1 of 0.82 with 10% of the modified test and 0.71 with 50% of the modified test. This is only possible if the model gathered sufficient information from the non label text in the input.

7 Conclusion and Future Work

Our chief motivation behind this work is the current pandemic and mandatory confinement worldwide. We believe that social media has become the prime mode of communication for people and has paved way for them to vent freely without judgement.

Our roadmap includes getting all our data annotated by mental health experts in order to verify our annotations. This would also assist us in creating a multi label dataset which is more representative of this problem when compared to a multi class one. We would also like to work on bettering our model on the behavioral tests. Our work involves two kinds of texts- long and short - both of which are common to the internet community. Hence, our work can easily be extended to many websites.

In conclusion, we believe that our work explores an interesting line of research where NLP is used to bridge the gap between virtual and real life of users and help those in need of medical attention.

References

Adrian Benton, Margaret Mitchell, and Dirk Hovy. 2017. Multi-task learning for mental health using social media text.

Glen Coppersmith, Mark Dredze, Craig Harman, Kristy Hollingshead, and Margaret Mitchell. 2015. Clpsych 2015 shared task: Depression and ptsd on twitter. In *Proceedings of the 2nd Workshop on Computational Linguistics and Clinical Psychology: From Linguistic Signal to Clinical Reality*. Association for Computational Linguistics.

Jacob Devlin, Ming-Wei Chang, Kenton Lee, and Kristina Toutanova. 2018. Bert: Pre-training of deep bidirectional transformers for language understanding.

George Gkotsis, Anika Oellrich, Sumithra Velupillai, Maria Liakata, Tim J. P. Hubbard, Richard J. B. Dobson, and Rina Dutta. 2017. Characterisation of mental health conditions in social media using informed deep learning. *Scientific Reports*, 7(1):1–10.

Sepp Hochreiter and Jurgen Schmidhuber. 1997. Long short-term memory.

Jina Kim, Jieon Lee, Eunil Park, and Jinyoung Han. 2020. A deep learning model for detecting mental illness from user content on social media. *Scientific Reports*, 10(1):1–6.

Diederik P. Kingma and Jimmy Ba. 2014. Adam: A method for stochastic optimization.

Yinhan Liu, Myle Ott, Naman Goyal, Jingfei Du, Mandar Joshi, Danqi Chen, Omer Levy, Mike Lewis, Luke Zettlemoyer, and Veselin Stoyanov. 2019. Roberta: A robustly optimized bert pretraining approach.

Ahmed Husseini Orabi, Prasadith Buddhitha, Mahmoud Husseini Orabi, and Diana Inkpen. 2018. Deep learning for depression detection of twitter users. In *Proceedings of the Fifth Workshop on Computational Linguistics and Clinical Psychology*. Association for Computational Linguistics.

Adam Paszke, Sam Gross, Francisco Massa, Adam Lerer, James Bradbury, Gregory Chanan, Trevor Killeen, Zeming Lin, Natalia Gimelshein, Luca Antiga, Alban Desmaison, Andreas Kopf, Edward Yang, Zachary DeVito, Martin Raison, Alykhan Tejani, Sasank Chilamkurthy, Benoit Steiner, Lu Fang, Junjie Bai, and Soumith Chintala. 2019. Pytorch: An imperative style, high-performance deep learning library. In *Advances in Neural Information Processing Systems 32*, pages 8024–8035. Curran Associates, Inc.

Marco Tulio Ribeiro, Tongshuang Wu, Carlos Guestrin, and Sameer Singh. 2020. Beyond accuracy: Behavioral testing of nlp models with checklist. In *Proceedings of the 58th Annual Meeting of the Association for Computational Linguistics*. Association for Computational Linguistics.

Ivan Sekulic and Michael Strube. 2019. Adapting deep learning methods for mental health prediction on social media. In *Proceedings of the 5th Workshop on Noisy User-generated Text (W-NUT 2019)*, pages 322–327, Hong Kong, China. Association for Computational Linguistics.

Nitish Srivastava, Geoffrey Hinton, Alex Krizhevsky, Ilya Sutskever, and Ruslan Salakhutdinov. 2014. Dropout: A simple way to prevent neural networks from overfitting. *J. Mach. Learn. Res.*, 15(1):1929–1958.

Ashish Vaswani, Noam Shazeer, Niki Parmar, Jakob Uszkoreit, Llion Jones, Aidan N Gomez, Ł ukasz Kaiser, and Illia Polosukhin. 2017. Attention is all you need. In I. Guyon, U. V. Luxburg, S. Bengio, H. Wallach, R. Fergus, S. Vishwanathan, and R. Garnett, editors, *Advances in Neural Information Processing Systems 30*, pages 5998–6008. Curran Associates, Inc.

Jason Wei and Kai Zou. 2019. Eda: Easy data augmentation techniques for boosting performance on text classification tasks. In *Proceedings of the 2019 Conference on Empirical Methods in Natural Language Processing*. Association for Computational Linguistics.

Thomas Wolf, Lysandre Debut, Victor Sanh, Julien Chaumond, Clement Delangue, Anthony Moi, Pierric Cistac, Tim Rault, Rémi Louf, Morgan Funtowicz, Joe Davison, Sam Shleifer, Patrick von Platen,

Clara Ma, Yacine Jernite, Julien Plu, Canwen Xu, Teven Le Scao, Sylvain Gugger, Mariama Drame, Quentin Lhoest, and Alexander M. Rush. 2019. Huggingface's transformers: State-of-the-art natural language processing. *ArXiv*, abs/1910.03771.

Ayah Zirikly, Philip Resnik, Özlem Uzuner, and Kristy Hollingshead. 2019. Clpsych 2019 shared task: Predicting the degree of suicide risk in reddit posts. In *Proceedings of the Sixth Workshop on Computational Linguistics and Clinical Psychology*. Association for Computational Linguistics.

Topic Modeling for Maternal Health Using Reddit

Shuang Gao
New York University
sg5963@nyu.edu

Shivani Pandya
John Hopkins University
spandya6@jhu.edu

Smisha Agarwal
John Hopkins University
sagarw23@jhu.edu

João Sedoc
New York University
jsedoc@stern.nyu.edu

Abstract

This paper applies topic modeling to understand maternal health topics, concerns, and questions expressed in online communities on social networking sites. We examine latent Dirichlet analysis (LDA) and two state-of-the-art methods: neural topic model with knowledge distillation (KD) and Embedded Topic Model (ETM) on maternal health texts collected from Reddit. The models are evaluated on topic quality and topic inference, using both auto-evaluation metrics and human assessment. We analyze a disconnect between automatic metrics and human evaluations. While LDA performs the best overall with the auto-evaluation metrics NPMI and Coherence, Neural Topic Model with Knowledge Distillation is favourable by expert evaluation. We also create a new partially expert annotated gold-standard maternal health topic modeling dataset for future research.[1]

1 Introduction

Evidence suggests that poor quality maternal and newborn care is responsible for nearly 60% of the estimated 5 million deaths each year globally (Kruk et al., 2018). This situation spouses urgent demands of artificial intelligence. AI has potential applications in automating aspects of service delivery such as basic counselling, thereby reducing the burden on health systems. In this work, we hope to help improving healthcare counselling AI by understanding maternal health related content. What are people concerning about regarding maternal health? And among state-of-the-art techniques, which model is the best at extracting and interpreting those highly-professional topics?

With the development of the online communities and social media, advanced text mining techniques are widely needed when applied to domains. It is necessary to understand a question in order to answer it, and the first step towards understanding the content of a document is to determine which topics that document addresses. Therefore, many successful conversational agents employ topic models to encourage more coherent dialogue (Baheti et al., 2018; Vlasov et al., 2019).

Topic models can provide legible and concise representations of both the entire corpus and individual documents because they represent topics as ranked word lists and documents regarding their probable topics. Therefore, while mainly applied to mining the text content, topic models can also classify the topics for new documents by inferring the latent topic distribution. In classical topic models like Latent Dirichlet Allocation (LDA), distributions over the latent variables are estimated with variational inference algorithms (EM) or Gibbs sampling (Blei et al., 2003). However, recently, with the development of deep learning methods, now there are neural topic models that incorporate additional information and leverage the variational autoencoder (VAE) framework for latent variable inference. Particularly, pre-trained transformer-based language models (e.g., BERT, (Devlin et al., 2019)) is employed to fine tune on a wide variety of NLP problems; some models combine the advantages of pre-trained transformers and topic models.

This work contributes the following:

- Provide an exemplary application of the three topic models (LDA, Neural topic model with knowledge distillation, and Embedding Topic Model) on texts in the maternity health domain.
- Evaluate the model performances from both quantitative and qualitative perspectives. The assessment may inspire ideas in metric design and model improvement.

[1] https://github.com/jsedoc/maternal_health_topics

Proceedings of the 12th International Workshop on Health Text Mining and Information Analysis, pages 69–76
April 19, 2021.

- Provide a new partially annotated dataset in the maternal healthcare domain for topic modelling.

2 Related Work and Background

There are a multitude of state-of-the-art topic models adapted from the LDA model. Neural topic models (Srivastava and Sutton, 2017) surpass traditional methods in architecture by using various forms of neural networks and can be applied to either labeled or unlabeled data. Hoyle et al. (2020) further improves neural topic models by knowledge distillation (KD). Another thread of innovations improve models in terms of semantic meaning of words, for example, Dieng et al. (2020) incorporates word embedding to topic models. Given that our dataset is unlabeled, this work focus on two state-of-the-art models, the KD model and the Embedding Topics Model (ETM), and compares their performances to LDA. We briefly summarize the details of LDA, KD and ETM in this section.

2.1 Latent Dirichlet Allocation

Latent Dirichlet Allocation (LDA) is a three-level hierarchical Bayesian model, in which each item of a collection is modeled as a finite mixture over an underlying set of topics. The topic probabilities provide an explicit representation of a document. LDA follows the generative process as below:

1. For each document d, draw the topic proportion θ_d from $Dir(\alpha)$;

2. For each word w_n in document d:

   (a) Choose a topic $z \sim Multinomial(\theta)$;

   (b) Choose a word w_n from p($w_n|z_n, \beta$), a multinomial probability conditioned on the topic z_n.

In our instance, we adopted MALLET (McCallum, 2002), a machine learning toolkit which implements LDA and conducts variational inference by Gibbs Sampling.

2.2 Neural Topic Model with Knowledge Distillation

Hoyle et al. (2020) combine neural topic modeling with knowledge distillation. It generates and stores the teacher logits (latent topics) $z^{teacher}$ for each document in the training set using DISTIL-BERT (Sanh et al., 2019) as a deterministic auto-encoder and training on document reconstruction as a decoder. Then it entails the framework of the neural topic model SCHOLAR but substitutes the reconstruction loss L_R with L_{KD}:

$$L_R = (w_d^{BoW})^T \log f(\theta_d, B)$$
$$L_{KD} = \lambda T^2 (w_d^{teacher})^T \log f(\theta_d, B; T) + (1 - \lambda) L_R$$
$$f(\theta_d, B) = \sigma(m + \theta_d^T B)$$

where $\sigma(\cdot)$ is the softmax function, $w_d^{teacher}$ is the result of teacher logits $z^{teacher}$ scaled by softmax temperature T, and $f(\cdot; T)$ is the scaled version of $f(\cdot)$. In doing so, the weights were updated to minimize the original reconstruction loss of SCHOLAR and the loss of reconstruction from teacher logits, which are pretrained and deterministic, so the knowledge contained in pretrained transformers is distilled to this student model. In this instance, we adopted the implementation of Hoyle et al. (2020).

2.3 Embedding Topics Model

Embedding Topics Model (ETM; Dieng et al., 2020) is also a generative probabilistic model that combines the useful properties of topic models and word embeddings. As a topic model, it discovers an interpretable latent semantic structure of the texts; as a word embedding, it provides a low-dimensional representation of the meaning of words. In contrast to LDA, the per-topic conditional probability of a term has a log-linear form that involves a low-dimensional representation of the vocabulary. Each term is represented by an embedding, and each topic is a point in that embedding space; the topic's distribution over terms is proportional to the exponentiated inner product of the topic's embedding and each term's embedding. ETM follows the generative process as below:

1. For each document d, draw the topic proportion θ_d from $LN(\alpha)$;
2. For each word w_n in document d:

   (a) Choose a topic $z \sim Multinomial(\theta)$;

   (b) Choose w_n from $softmax(\rho^T \alpha_z)$, a multinomial probability computed by the inner product of word embedding and the embedding of topic z.

In this instance, we adopted the implementation of Dieng et al. (2020).

3 Experimental Details

3.1 Data

3.1.1 Collection

For our project, we deploy data from the popular site Reddit spanning between December 2005 and April 2019 because of the abundance of data in almost any given topic due to the site's specialized "subreddits". The public data on Reddit is accessible, which provides convenience to researchers, and the specialized channel create an online community about maternal healthcare, making it easier to collect and filter a domain corpus. To make the data set more manageable and remove unwanted data, we eliminated all subreddits not in a precompiled list of relevant subreddits chosen by a healthcare professional. Finally we end up with 24k comments for training and 6k comments as hold-out samples. Table 1 is a statistical description of the corpus, and we release this maternal health corpus on the github.

# documents	30647
Avg length per doc	81.51
# Vocabulary (over 3 counts)	11537
# Vocabulary (over 100 counts)	1880
# Vocabulary (over 1000 counts)	336
# Vocabulary (over 10000 counts)	37

Table 1: Statistics description of the collected corpus

3.1.2 Preprocessing

Before applying the models on this dataset, we preprocess the data for each model to prepare valid tokens or word embeddings. For LDA and the neural topic model with knowledge distillation (KD), we preprocess the data by removing common stop words in MALLET stoplist. LDA uses MALLET's built-in tokenizer, and the other two models tokenize the text by single words and convert the word lists to bag of words. Except for the KD model which requires pretraining with the teacher model, ETM also requires pretraining the word embedding on the training documents.

3.2 Metrics

3.2.1 Coherence

Mimno et al. (2011) proposed coherence to measure the quality of topics. Coherence measures topic word co-occurrence across documents to detect low quality topics, and show it correlates with expert topic annotations. The Coherence of topic t is defined as

$$coherence(t) := \sum_{i=1}^{N-1} \sum_{j=i+1}^{N} \log \frac{D(v_i^t, v_j^t)}{D(v_i^t)},$$

$D(x)$ represents the number of documents word x appears in, and $D(x, y)$ represents the number of documents x and y co-appear in. Therefore, if the words in the same topic always appear in the same documents, this topic is coherent, and the score approaches zero. Otherwise, if the words always appear in different documents, the coherence score approaches negative infinity.

3.2.2 NPMI

First introduced by Bouma (2009), Normalized Pointwise Mutual Information (NPMI) is adapted from PMI. Newman et al. (2010) shows that the auto-evaluation of topic-semantic coherence using PMI is highly correlated with human evaluation, and NPMI has been widely used as a quantitative measurement of topic quality (Aletras and Stevenson, 2013; Hoyle et al., 2020). The NPMI of topic t is defined as

$$NPMI(t) := \sum_{i=1}^{N-1} \sum_{j=i+1}^{N} \frac{\log \frac{P(v_i^t, v_j^t)}{P(v_i^t)P(v_j^t)}}{-\log P(v_i^t, v_j^t)}.$$

3.2.3 Topic Uniqueness

We also use Topic Uniqueness (TU) as a quantitative measurement. As the coherence and NPMI measure how well the top words within a topic share similar context, TU measures to what extend the different topics overlap. The TU for topic t is defined as

$$TU(t) := \frac{1}{N} \sum_{i=1}^{N} \frac{1}{count(v_i^t)}$$

where $count(v_i^t)$ is the total number of times the word v_i^t appears in top N words across all topics. If $TU(t)$ equals to 1, then all top words of topic t don't appear in top words of any other topics.

4 Experimental Results

4.1 Automatic Evaluation

We train each model under 10, 25, 50, 100, and 200 topic settings, and compare them with respect to the three metrics. Table 6, Table 7 and Table 8 in Appendix A show the performances of different models in terms of Coherence, NPMI and Topic Uniqueness respectively.

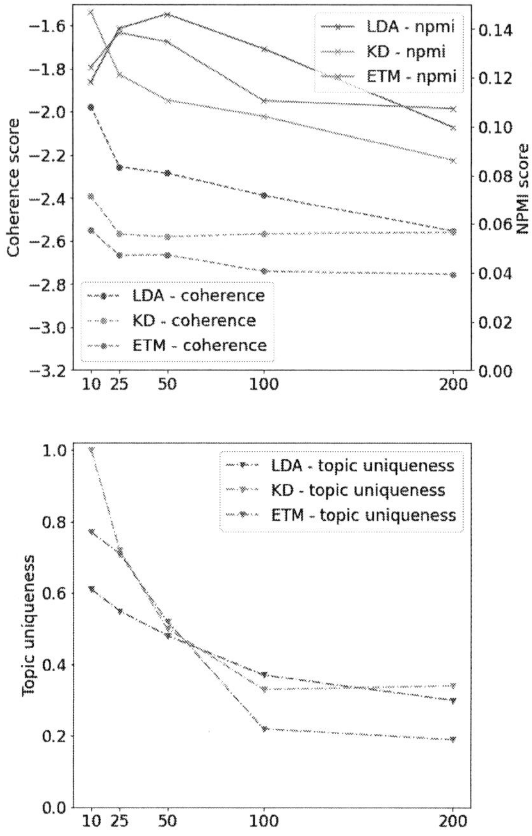

Figure 1: Performances of different models. The top plot shows the coherence scores and the NPMI scores, and the bottom plot shows the topic uniqueness.

4.2 Human Assessment

Two experts in public health identified topics generated by the 10-topic and 25-topic models. By reading through the top words for each topic, one expert annotates effective words and summarize the topic, and the other checks the annotation and topic names, and makes complement. They make an agreement on the final assessment.

ETM dismisses more topics than the other two models. When looking for 10 topics with each model, there are a total of 10 topics identified from all topics generated by the three models, and only 5 of them are identified in all of the three models: "Pregnancy", "Abortion", "Vaginal health", "Menstruation", "Labor and delivery". Besides that, only the KD model generated topics which are identified as "Penile health", "Breastfeeding" and "pregnancy symptoms". ETM also generated the least identifiable topics when looking for 25 topics. Out of the 13 topics identified, ETM dismisses "Sex", "Maternity and paternity leave" and "Stis", and it had 4 topics that cannot be identified as maternity-related.

Moreover, the KD model has the most representative top words. The expert was asked to annotate related words among the 20 top words within each topic, and the KD model achieves 17.8 and 15.7 words on average for the 10-topic and 25-topic models respectively, much more than LDA and ETM, as shown in Table 4

4.3 Inference

To infer the topics for the new documents held-out from the training data, we sampled 10 comments from the dev set and inferred their topics one by one. According to the performance reported above, we selected LDA 25 and KD 25 to make the inference. Among the 10 comments displayed in Appendix B, KD had better topic classification. LDA and KD misclassified comment 1 as "penile health" and "vaginal health" respectively, while the comment is related to "pregnancy". Comment 6 was also misclassified by both. Comment 8 was misclassified as "pregnancy" by LDA but classified as "abortion" correctly by KD. Comment 9 was classified as "pregnancy" by LDA, which was correct, but KD more accurately classification as "pregnancy risk". All other comments are correctly classified by both models.

5 Discussion

Measurement Incompatibility There is an incompatibility between human assessment and the auto-evaluation metrics for topic quality. In terms of the auto-evaluation metric Coherence, LDA always performs the best, which is inconsistent with the human assessment. In terms of NPMI, KD dominates in the 10-topic setting, but LDA still performs better in the 25-topic, 50-topic, and 100-topic settings. In terms of Topic Uniqueness, KD is the best in 10-topic, 25-topic setting, which is the same as human evaluation.

Topic Repetitiveness Although fixed values are set for the number of topics when training on the corpus, fewer topics could be identified from the generated word lists because of the overlapping among the lists. For example, when generating 10 topics, KD has 9 topics identified by the expert, while topic 3 and topic 5 are categorized as "vaginal health". For ETM, there are more overlapped generated lists: topic "pregnancy", "vaginal health", "menstruation", and "labor and delivery" all include 2 generated lists.

Method	Topic Words
LDA	**life abortion child** would **fetus** think woman right human people person mother **choice** body **baby** pro women want **abortions** one
KD	**murder alive** human **fetus personhood rights zygote** argument **fetuses womb** begins **killing survive pro clump embryo viable debate development life**
ETM	**abortion life child fetus** people woman human women point person body make mother **abortions baby** medical **support** agree **choice** circumcision

Table 2: Topic words associated with the topic labelled as abortion. Bolded words are marked as topic relevant.

	LDA	KD	ETM
n/a	1 / 1	- / 1	- / 4
pregnancy	1 / 5	1 / 3	2 / 4
birth control	1 / 3	1 / 5	- / 1
abortion	1 / 2	1 / 2	1 / 3
menstruation	2 / 3	1 / 2	2 / 3
vaginal health	1 / 2	2 / 3	2 / 3
penile health	- / 1	1 / 1	- / 1
sex	1 / 1	- / 1	1 / -
breastfeeding	- / 1	1 / 1	- / 1
pregnancy symptoms	- / 1	1 / -	- / 2
labor and delivery	2 / 2	1 / 3	2 / 2
pregnancy risks	- / 1	- / 1	- / 1
maternity leave	- / 1	- / 1	- / -
stis	- / 1	- / 1	- / -

Table 3: Identified topics counts for 10-topics and 25-topic models (separated by '/')

	LDA	KD	ETM
10-topic	6.7	**17.8**	8.7
25-topic	9.5	**15.7**	10.6

Table 4: Average number of effective top words on topics

Word Frequency of Top Words As shown in Tables 2 and 5, LDA and ETM include more frequently appeared words such as "feel", "period" and "life" in top word lists, the top words of each topic generated by KD is professional and of low-frequency.

6 Conclusion

We applied and compared three different topic models on the texts related to maternal health and discovered topics with high-quality. By both auto-evaluation and human assessment, we evaluated the topic quality generated by different models and observed incompatibility among metrics. Taking the human assessment as a gold standard, we would recommend applying the neural topic model with knowledge distillation, because it tends to rank professional words that are unique to one topic as top words, and it covers more diverse topics when asked to generate a fixed number of topics. Regarding topic inference on new documents, both LDA and the knowledge-distilled model perform well.

	LDA	KD	ETM
10-topic	0.1102	0.0132	0.0558
25-topic	0.0881	0.0280	0.0441
50-topic	0.0805	0.0275	0.0424
100-topic	0.0747	0.0250	0.0407
200-topic	0.0662	0.0240	0.0432

Table 5: Average word frequencies for topics generated. The frequency of a word w is represented with the proportion of documents which contain the word w.

References

Nikolaos Aletras and Mark Stevenson. 2013. Evaluating topic coherence using distributional semantics. *In Proceedings of the 10th International Conference on Computational Semantics*, pages 13–22.

Ashutosh Baheti, Alan Ritter, Jiwei Li, and Bill Dolan. 2018. Generating more interesting responses in neural conversation models with distributional constraints. In *Proceedings of the 2018 Conference on Empirical Methods in Natural Language Processing*, pages 3970–3980, Brussels, Belgium. Association for Computational Linguistics.

David M. Blei, Andrew Y. Ng, and Michael I. Jordan. 2003. Latent dirichlet allocation. *Journal of machine Learning research*, 3(Jan):993–1022.

Gerlof Bouma. 2009. Normalized (pointwise) mutual information in collocation extraction. *In Proceedings of the Biennial GSCL Conference*, 1:31–40.

Jacob Devlin, Ming-Wei Chang, Kenton Lee, and Kristina Toutanova. 2019. Bert: pre-training of deep bidirectional transformers for language understanding. *In Proceedings of the 2019 Conference of the*

North American Chapter of the Association for Computational Linguistics: Human Language Technologies, 1:4171–4186.

Adji B. Dieng, Francisco J.R. Ruiz, and David M. Blei. 2020. Topic modeling in embedding spaces. *Transactions of the Association for Computational Linguistics*, 3:439–453.

Alexander Hoyle, Pranav Goel, and Philip Resnik. 2020. Improving neural topic models using knowledge distillation. *In Proceedings of the 2020 Conference on Empirical Methods in Natural Language Processing*, pages 1752–1771.

Margaret E. Kruk, Anna D. Gage, Catherine Arsenault, Keely Jordan, Hannah H. Leslie, Sanam Roder-DeWan, Olusoji Adeyi, Pierre Barker, Bernadette Daelmans, Svetlana V. Doubova, Mike English, Ezequiel García Elorrio, Frederico Guanais, Oye Gureje, Lisa R. Hirschhorn, Lixin Jiang, Edward Kelley, Ephrem Tekle Lemango, Jerker Liljestrand, Address Malata, Tanya Marchant, Malebona Precious Matsoso, John G. Meara, Manoj Mohanan, Youssoupha Ndiaye, Ole F. Norheim, K. Srinath Reddy, Alexander K. Rowe, Joshua A. Salomon, Gagan Thapa, Nana A. Y. Twum-Danso, and Muhammad Pate. 2018. High-quality health systems in the sustainable development goals era: time for a revolution. *The Lancet Global Health*, 6(11):e1196–e1252.

Andrew Kachites McCallum. 2002. Mallet: A machine learning for language toolkit. *Http://mallet.cs.umass.edu*.

David Mimno, Hanna M. Wallach, Edmund Talley, Miriam Leenders, and Andrew McCallum. 2011. Optimizing semantic coherence in topic models. *In Proceedings of the 2011 Conference on Empirical Methods in Natural Language Processing*, pages 262–272.

David Newman, Jey Han Lau, Karl Grieser, and Timothy Baldwin. 2010. Automatic evaluation of topic coherence. *In Human language technologies: The 2010 annual conference of the North American chapter of the association for computational linguistics*, pages 100–108.

Victor Sanh, Lysandre Debut, Julien Chaumond, and Thomas Wolf. 2019. Distilbert, a distilled version of bert: smaller, faster, cheaper and lighter. *arXiv preprint arXiv:1910.01108*.

Akash Srivastava and Charles Sutton. 2017. Autoencoding variational inference for topic models. *arXiv preprint arXiv:1703.01488*.

Vladimir Vlasov, Johannes E.M. Mosig, and Alan Nichol. 2019. Dialogue transformers. *arXiv preprint arXiv:1910.00486*.

	LDA	KD	ETM
10-topic	**-1.9756**	-2.3899	-2.5505
25-topic	**-2.2545**	-2.5681	-2.6672
50-topic	**-2.2841**	-2.5807	-2.6649
100-topic	**-2.3867**	-2.5663	-2.7402
200-topic	**-2.5528**	-2.5592	-2.7535

Table 6: Coherence for topics generated

	LDA	KD	ETM
10-topic	0.1181	**0.1469**	0.1243
25-topic	**0.1401**	0.1212	0.1384
50-topic	**0.1460**	0.1107	0.1347
100-topic	**0.13190**	0.1041	0.1105
200-topic	0.0997	0.0862	**0.1073**

Table 7: NPMI for topics generated

A Appendix: Experiment Results

The experiment results are shown in Table 6, Table 7 and Table 8

We trained the LDA model with 4 CPU cores and trained the other two model with a single GPU core. Table 9 shows that LDA implemented by Mallet is quite faster than the other two models, and KD is the most time-consuming one because of the pre-training of teacher model. But once the logits of teacher model has been saved, it is fast to train the student models with various number of topics.

B Appendix: Examples for Inference

Comment 1 I was waiting to see if someone brought this up! Alcohol and tobacco are not the same. Moderate alcohol use (one drink a day or less) hasn't been shown to be harmful to a pregnancy. ANY tobacco use is harmful to a pregnancy. I would feel much more comfortable refusing to serve hooka to a pregnant woman that refusing to serve alcohol. It's actually been shown that moderate alcohol consumption can be beneficial (-drinking-pregnant-women.html)!

	LDA	KD	ETM
10-topic	0.61	**1.00**	0.77
25-topic	0.55	**0.72**	0.71
50-topic	0.48	0.50	**0.52**
100-topic	**0.37**	0.33	0.34
200-topic	**0.30**	0.22	0.19

Table 8: Topic Uniqueness for topics generated

	LDA	KD	ETM
pre-train		teacher model 6409 s	embedding 10 s
10-topic	111 s	364 s	1112 s
25-topic	131 s	186 s	1121 s
50-topic	148 s	236 s	1110 s
100-topic	167 s	340 s	1062 s
200-topic	204 s	547 s	1086 s

Table 9: Training time for different models

(LDA - penile health, KD - vaginal health)

Comment 2 So will you be back for a follow up pap in a few months or getting the colposcopy right away? A lot of these clear up on their own. Try not to panic!
(LDA - vaginal health, KD - vaginal health)

Comment 3 I was on Loestrin24Fe for about a year but at my February yearly gyno appointment, I asked my doctor about a pill to give me less periods because mine, like yours, were *brutal.* He switched me to Lo Loestrin Fe and I haven't had a period since! It's AWESOME. I've also noticed that I am not as emotional as I was on L24Fe. Don't get me wrong - For about the first month, I was super sensitive to everything anyone said to me. But since then, I've been fine. My sex drive is sky high as always - but I don't think ANYTHING could kill my libido! Haha. I have not gained any weight since starting this pill. Hope this helps!
(LDA - birth control, KD - birth control)

Comment 4 Personally, I don't remember distinctly having that feeling (I had an epidural), but I obviously would have no idea if I would have otherwise. I think it was mostly just because I was way too exhausted to feel anything but relief (over 30 hours of labor, and no sleep), also if anything, I think the pitocin interfered with it since that's synthetic oxytocin. However, I totally felt an instant bond with my son and would not put him down those first few hours, I didn't even go to sleep.
(LDA - labor and delivery, KD - labor and delivery)

Comment 5 The only reason I ever heard about not being able to eat is in case of an emergency c-section where you are put under general anesthesia. I guess it depends on where you are and what they do, but general anesthesia is not the common practice for emergency c-section anymore. If you were already having an epidural they would work

with that, or they would give you a spinal in an emergency. That way you are still awake for the operation and there is minimal risk. When getting an epidural, a common side effect is a drop in blood pressure so an IV of fluids is introduced to offset this. That would be one reason to have IV fluids during labour. I plan on eating and drinking during my labour because, like you,I think how the hell are you supposed to be able to push out a baby when you have been starved for a day! Thankfully my midwives are of the same mind. So I will be packing snacks and a few sandwiches to bring with me for sure.
(LDA - labor and delivery, KD - labor and delivery)

Comment 6 Colon cancer can be *prevented* by regular colonoscopies. They remove precancerous polyps while they're up there so you won't ever develop it in the first place. It honestly has to be one of the most preventable cancers, and too many people know nothing about it. If you're over 40, get the pooper scope! It's a good decision.
(LDA - vaginal health, KD - vaginal health)

Comment 7 They make some very small models (check a comparison chart for the actual sizes), and keep in mind that cups are easier to insert than tampons for a lot of people (like me), because tampons are dry. It's pretty easy to moisten cups with a bit of water (or lube, even), since you're in the bathroom anyway. [Here's the latest comparison chart I could find.] For me, they've been a lifesafer since I can't use tampons, pads and my sensitive skin don't get along, and because I have a very heavy flow. They're not for everyone, but they're definitely the most comfortable alternative I've tried.
(LDA - menstruation, KD - menstruation)

Comment 8 Well, kinda yeah. there are other ways to keep yourself from having a baby besides abortions, and tbh, abstinence isn't the end of the world, lol. It isn't fair of anyone to think its okay to destroy the potential of an unborn child because they simply don't want to deal with the consequences of a certain lifestyle. Change your lifestyle, or use better protection, or have the child anyway, but don't just write off the life of a fetus like it is worth nothing, because it isn't.
(LDA - pregnancy, KD - abortion)

Comment 9 My cycles were off, so I had a decent wait. I went to the health dept to get confirmation and health insurance, then scheduled a drs appointment which was a week or two away, the doctor guessed I was about 6-7 weeks, though my lmp said I was 14 weeks because he couldn't find a heartbeat with Doppler and he was feeling for my uterus. Then I was called the next day with a date for an ultrasound about two weeks away and was told to go back to get the rest of my blood drawn, as they only took one vial the day before. When I should have been 16 weeks, I was 12 weeks, as it showed on the ultrasound. Ultimately, you may as well wait. Waiting is something you deal with all of pregnancy, and your wait is really short, so if anything, I'd wait for the appointment and express your angst to find out how far along she is in hopes of getting things moving.
(LDA - pregnancy, KD - pregnancy risks)

Comment 10 I've had my Mirena for nearly 5 years (getting it replaced this summer) and I'm one of the lucky ones where insertion didn't hurt at all and compared to when I was on the pill, my sex drive went waaaaay up. Not sure if it returned to "normal" or not since I had been on the pill for years and years before I became sexually active so I have nothing to compare it to, but my boyfriend is a lot happier now :p I also haven't had a "real" period in about 4 years, as compared to every two to three weeks even on the pill. Needless to say, I'm a huge personal Mirena fan :)
(LDA - birth control, KD - birth control)

FuzzyBIO: A Proposal for Fuzzy Representation of Discontinuous Entities

Anne Dirkson
LIACS, Leiden University
Niels Bohrweg 1, 2333 CA,
Leiden, the Netherlands

Suzan Verberne
LIACS, Leiden University
Niels Bohrweg 1, 2333 CA
Leiden, the Netherlands

Wessel Kraaij
LIACS, Leiden University
Niels Bohrweg 1, 2333 CA
Leiden, the Netherlands

{a.r.dirkson, s.verberne, w.kraaij}@liacs.leidenuniv.nl

Abstract

Discontinuous entities pose a challenge to named entity recognition (NER). These phenomena occur commonly in the biomedical domain. As a solution, expansions of the BIO representation scheme that can handle these entity types are commonly used (i.e. BIOHD). However, the extra tag types make the NER task more difficult to learn. In this paper we propose an alternative; a fuzzy continuous BIO scheme (FuzzyBIO). We focus on the task of Adverse Drug Response extraction and normalization to compare FuzzyBIO to BIOHD. We find that FuzzyBIO improves recall of NER for two of three data sets and results in a higher percentage of correctly identified disjoint and composite entities for all data sets. Using FuzzyBIO also improves end-to-end performance for continuous and composite entities in two of three data sets. Since FuzzyBIO improves performance for some data sets and the conversion from BIOHD to FuzzyBIO is straightforward, we recommend investigating which is more effective for any data set containing discontinuous entities.

1 Introduction

Adverse Drug Reactions (ADRs), harmful reactions that result from the intake of medication, pose a major health concern (World Health Organisation, 2006). Due to the limitations of clinical trials on the one hand (Shenoy and Harugeri, 2015) and reporting systems after release on the market on the other hand (Hazell and Shakir, 2006), many ADRs remain undiscovered. Therefore, both social media and clinical reports are being explored by the research community as alternative information sources for the semi-automatic discovery of ADRs (Lardon et al., 2015; Sarker et al., 2015).

One particular challenge for the extraction of ADRs from text is the presence of discontinuous entities. These can be either composite entities (i.e. some words belong to multiple entities), such as '*lack of sleep* and *appetite*', or disjoint entities (i.e. split entities), such as '*eyes* are feeling *dry*'. These phenomena occur more commonly in the clinical than general domain. In fact, Tang et al. (2015) reported that discontinuous mentions in clinical text account for about 10% of all ADR mentions. None of the traditional versions of the BIO representation scheme (B: beginning of entity, I: inside entity and O: outside entity) or common extensions such as IOBES (E: end of entity, S: singleton entity)[1] were designed to handle such mentions (Pradhan et al., 2014). Therefore, Tang et al. (2015) proposed extending the BIO scheme with two additional tags: the 'H' for words shared by multiple mentions and 'D' for parts of discontinuous mentions not shared by other mentions. This resulted in four new tag types (HB-, HI-, DB- and DI-). Their BIOHD representation was broadly adopted by the community (Karimi et al., 2015; Zolnoori et al., 2019; Si et al., 2019). Table 1 shows examples of concepts represented with the BIOHD scheme.

Although the BIOHD scheme allows for precise representation of entities, the extra tag types make the task more difficult for models to learn. Straightforward BIO rules such as 'an entity always starts with a B' are no longer valid under the BIOHD scheme. In this paper we argue that a more simple BIO representation in which discontinuous entities are transformed into continuous sequences by including all non-entity tokens in between would improve ADR extraction by being easier to learn and reintroducing these straightforward rules. We coin this representation FuzzyBIO. Some examples of entities represented with BIOHD and FuzzyBIO can be seen in Table 1.

Aside from improving extraction, using FuzzyBIO instead of BIOHD may also improve sub-

[1]This scheme is also called BIOES, BILOU or BMEWO

Proceedings of the 12th International Workshop on Health Text Mining and Information Analysis, pages 77–82
April 19, 2021. ©2021 Association for Computational Linguistics

Sentence 1	**Muscles**	are	constantly	**quivering**	!				
BIOHD	DB	O	O	DI	O				
FuzzyBIO	B	I	I	I	O				
Sentence 2	I	have	**pain**	**in**	**my**	**hands**	and	**upper**	**arms**
BIOHD	O	O	HB	HI	HI	DB	O	DB	DI
FuzzyBIO	O	O	B	I	I	I	I	I	I

Table 1: Examples of discontinuous disjoint (sentence 1) and composite (sentence 2) ADR mentions represented by the BIOHD and FuzzyBIO schemes.

sequent concept normalization, in which ADRs are linked to standardized medical concepts (e.g. 'can't fall asleep' to the concept 'insomnia' with concept identifier 193462001 in the medical ontology SNOMED-CT). This step is essential for aggregating and thus quantifying the prevalence of ADR. As current normalization methods are mostly hampered by errors made during extraction (Weissenbacher et al., 2019), this step may also benefit from simplification of the representation scheme.

Thus, we address two research questions:

RQ1 To what extent can fuzzy continuous representation of discontinuous entities improve NER of ADR? (intrinsic evaluation)

RQ2 To what extent can this fuzzy representation benefit end-to-end ADR extraction? (extrinsic evaluation)

In this paper we present FuzzyBIO, a fuzzy continuous BIO representation of discontinuous entities. Moreover, we show it is beneficial for end-to-end ADR discovery. Our representation is applicable to other domains as well. We release our code for the purpose of follow-up research.[2]

2 Related Work

The first shared task to deal with discontinuous medical entities was SemEval 2014 Task 7 (Pradhan et al., 2014). Prior to this task, discontinuous entities were often excluded (e.g. Uzuner et al. (2011)) or each part was represented as a separate continuous entity and later reassembled (Metke-Jimenez and Karimi, 2015). Various representations were proposed but the only one able to distinguish between those that share a head word (i.e composite entities) and those that do not (e.g. disjoint entities) was the BIOHD scheme (Zhang et al., 2014).

This scheme was later analysed in more detail and compared to two baseline approaches: (1) ignoring all discontinuous entities and (2) representing separate parts of discontinuous entities as individual entities (Tang et al., 2015). In comparison to the baseline approaches, the BIOHD scheme could improve recognition of both discontinuous but also continuous entities, likely due to its ability to distinguish between the two.

Tang et al. (2015) also proposed a further extension (BIOHD1234) in which numbers were added to refer to which entity a non-head ('D') entity should be combined with[3], effectively expanding the scheme from 7 to 13 tags. This representation was able to outperform BIOHD due to its ability to correctly represent multiple discontinuous entities and discontinuous entities with more than one non-head part. However, as neither BIOHD nor BIOHD1234 could handle multiple head entities in one sentence, Tang et al. (2018) proposed a multi-label BIO representation in which tokens can be labeled with more than one tag, and each tag corresponds to one entity. For NER of adverse drug responses, this novel representation managed to outperform BIOHD. Similarly, Shang et al. (2020) allowed for multiple labels per token for extracting disorders from scientific articles.

Despite its limitations (Tang et al., 2018), the BIOHD scheme is commonly adopted (Si et al., 2019; Karimi et al., 2015; Zolnoori et al., 2019). We propose an alternative, simpler representation scheme that could improve extraction by being easier to learn.

3 Methods

3.1 The FuzzyBIO representation scheme

As displayed in Table 1, FuzzyBIO transforms discontinuous into continuous entities by annotating

[2]Code is available at: `https://github.com/AnneDirkson/FuzzyBIO`

[3]1 and 2 denote nearest head and non-head entity on the left, and 3 and 4 denote nearest head and non-head entity on the right

all tokens in between.[4] Composite entities are combined if they share an entity head. We realise this compresses two separate entities into one (e.g. the entities 'pain in my hands' and 'pain in my upper arms' in Table 1). However, this does not pose a problem to normalization, as the state-of-the-art normalization method (Sung et al., 2020) includes heuristic rules to split composite entities prior to normalization.

3.2 Named Entity Recognition of ADR

For the NER task itself, we opt for distilBERT (base-cased), a lighter more computationally efficient version of BERT (Sanh et al., 2019). We use a one-cycle learning rate (LR) policy (Smith, 2018) with a maximum LR of 0.01. For each fold in the 10-fold cross-validation (CV), we select either 3 or 4 epochs based on the validation data. We use the Huggingface implementation (Wolf et al., 2019) with the wrapper ktrain (Maiya, 2020) to train our models with the initialization seed set to 1.

3.3 Concept normalization of ADR

For normalization, we use the state-of-the-art BioSyn method with default parameters (Sung et al., 2020; Tutubalina et al., 2020). It is possible to provide composite entities as input, as this method splits composite entities prior to normalization using the heuristics by Souza and Ng (2015).

Our target ontology is SNOMED-CT[5]. As SNOMED-CT is too extensive for our purpose, we aim to map SNOMED concepts in our training data to a curated subset of SNOMED, the CORE Problem List Subset[6], before training the normalization model. If there is a direct mapping in the community based mappings in BioPortal (Noy et al., 2008) between the original concept and a CORE concept or the parent of the concept is in the CORE (e.g. 'moderate anxiety' to 'anxiety'), we map the mention to the respective CORE concept. We include all concepts of the CORE subset and all concepts that could not be mapped to a CORE concept in the data as candidates. Synonyms for each concept are retrieved from the community based mappings in BioPortal (Noy et al., 2008) using the REST API and from the UMLS using pymedtermino (Lamy et al., 2015).

Entities	CADEC	PsyTAR	CLEF
Continuous	5.360	4.508	16.261
Discontinuous			
– Disjoint	100	225	909
– Composite	828	70	286

Table 2: Size of data sets.

3.4 Evaluation

For evaluating the NER models on a token level, our metrics are lenient and ignore the prefixes (B- I- H- D-). Additionally, we evaluate performance on an entity level. Following Magge et al. (2020), an entity is considered a true positive if any part of the annotated ADR text is correctly identified (i.e. overlaps with the predicted ADR text). We evaluate the end-to-end performance by calculating how many entities were both extracted during NER and normalized to the correct SNOMED-CT concept.

4 Data

We use two data sets of social media posts annotated for ADR: CADEC (Karimi et al., 2015) and PsyTAR (Zolnoori et al., 2019). The former was also used by Tang et al. (2018). Both contain posts from medical fora on AskaPatient.com. Additionally, we used a data set of clinical records annotated for disorder mentions, namely the SemEval 2014 Task 7 data (Pradhan et al., 2014) that builds on the CLEF eHealth 2013 corpus used by Tang et al. (2015). See Table 2 for more details. Data sets were split into 10 folds stratified on the presence of ADRs. For PsyTAR, we chose sentences as units as they were annotated separately. For the CLEF data set, each document was split into sequences of 5 sentences for the NER task, because of memory restrictions on the input length for BERT models.

5 Results

5.1 Intrinsic evaluation

As can be seen in Table 3, the FuzzyBIO scheme improves recall for two of three data sets, namely for CADEC (+0.29) and CLEF (+0.07), at a cost to precision (-0.24 and -0.1). For these data sets, using FuzzyBIO also leads to a higher percentage of correctly identified entities for both continuous (+1.1 and +0.6) and discontinuous entities (+3.5

[4] In our data, we did not find any cases where an entity was lost because it was in between two parts of another discontinuous entity. Nonetheless, this is theoretically possible and poses a potential limitation.

[5] https://www.nlm.nih.gov/healthit/snomedct/index.html

[6] https://www.nlm.nih.gov/research/umls/Snomed/core_subset.html

Data	Scheme	Token level Micro F_1	P	R	Entity level recall Continuous	Disjoint	Composite
CADEC	BIOHD	0.586	**0.636**	0.555	42.0%	55.4%	64.6%
	FuzzyBIO	**0.596**	0.612	**0.584**	**43.1%**	**58.9%**	**65.2%**
PsyTAR	BIOHD	**0.771**	**0.751**	**0.797**	**87.4%**	83.8%	79.5%
	FuzzyBIO	0.762	0.747	0.780	84.9%	**84.4%**	**87.5%**
CLEF	BIOHD	**0.312**	**0.286**	0.345	35.6%	52.8%	69.1%
	FuzzyBIO	0.309	0.276	**0.352**	**36.2%**	**55.2%**	**76.6%**

Table 3: Intrinsic evaluation of NER. Results are the average of a 10 fold CV.

and +3.6 for disjoint and +0.6 and +0.7 for composite entities). For the remaining data set (PsyTAR), overall NER performance is negatively affected by using FuzzyBIO (-0.09) and continuous entities are missed more often (-2.5). Nonetheless, also for this data set discontinuous entities are extracted correctly more often (+0.8 and +8.0) when using FuzzyBIO instead of the BIOHD scheme.

5.2 Extrinsic evaluation

As can be seen in Table 4, using the FuzzyBIO scheme improves end-to-end performance for continuous and composite entities in two of three data sets, namely the CADEC (+0.4 and +1.3) and CLEF data (+0.4 and +5.7). In contrast, the end-to-end performance for disjoint entities is decreased (-15.5 and -21.0) for these data sets despite initial gains during NER. In the remaining data set (PsyTAR), the percentage of correctly identified entities after normalization is lower for all entity types when using the FuzzyBIO instead of the BIOHD scheme.

Data	Scheme	Entity level recall Continuous	Disjoint	Composite
CADEC	BIOHD	23.9%	**35.9%**	21.2%
	FuzzyBIO	**24.3%**	20.4%	**22.5%**
PsyTAR	BIOHD	**43.6%**	**26.1%**	**10.6%**
	FuzzyBIO	42.8%	25.0%	7.5%
CLEF	BIOHD	21.7%	**25.8%**	26.5%
	FuzzyBIO	**22.1%**	4.8%	**32.2%**

Table 4: Extrinsic evaluation of ADR extraction. Results are the average of a 10 fold CV.

6 Discussion

In answer to RQ1, we find that the FuzzyBIO scheme benefits overall recall during NER for two of the three data sets. For these data sets, it also leads to a higher percentage of correctly identified entities, both continuous and discontinuous. For the third data set (PsyTAR), more discontinuous entities are extracted correctly when using FuzzyBIO compared to the BIOHD scheme. However, more continuous entities are missed. In answer to RQ2, we find that for the same two data sets (CADEC and CLEF) the end-to-end ADR extraction is improved for continuous and composite entities. However, for the remaining data set (PsyTAR) the end-to-end performance is lower for all entity types.

We believe that the difference between PsyTAR and the other data sets may be related to either the low number of discontinuous entities or the low number of composite entities in the PsyTAR data, which may have hindered the training of an NER model for these entity types. An alternative explanation is that FuzzyBIO is less beneficial for easier NER tasks: The initial NER performance with BIOHD is far higher for PsyTAR (F_1 of 0.771) than for the other data sets (F_1 of 0.586 and 0.312). The difference between PsyTAR and the other data sets is unlikely to be related to the relative percentage of discontinuous entities, as this is similar to that of the CLEF data (5.8 vs 6.2%), or the nature of the data, as CADEC contains forum posts from the same website.

Another result that stands out is the lower end-to-end performance for disjoint entities when using the FuzzyBIO scheme despite initial gains in the extraction of disjoint entities for all data sets. We suspect that normalization of these entities is made more challenging by the words in between the disjoint parts of the entity that are now included in the extracted entity. Therefore, in future work, we plan to investigate post-processing steps such as the removal of stop words which may improve the normalization of the more noisy disjoint entities represented with FuzzyBIO. As our representation is applicable for representing any type of discontinuous entity, future work may also include testing FuzzyBIO in other domains.

Although the improvement in NER is comparable for disjoint entities in medical records and user-generated content, the negative impact on normalization of disjoint entities is far stronger for the medical records. One might expect the normalization to decrease more strongly if more non-entity words (i.e. more noise) were included, but the median amount of non-entity words included in the disjoint entities is equal for all data sets (on average 1 word is added). Manual analysis also does not reveal a difference between the *type* of non-entity word included; They appear to mostly be stopwords. Thus, the most likely explanation for this difference is that the training examples for normalization from the user-generated data are already more noisy than their counterparts from the medical records. Consequently, the normalization algorithm for user-generated data might be better at dealing with noise. Future work could investigate whether training with the noisy examples instead of the original entities would be beneficial.

FuzzyBIO appears to be more beneficial for end-to-end extraction of composite entities in the medical records (+5.7) than in the user-generated data (+1.3 and -3.1). However, the number of non-entity words that is included is not lower for the medical records (median of 2 words added) compared to the user-generated data (median of 1 for CADEC and 3 for PsyTAR). Thus, this difference does not appear to be due to an increase in the fuzziness of the entities.

We also find some support for our hypothesis that the BIOHD representation makes the NER task more difficult for BERT models to learn than the FuzzyBIO representation. Overall the BERT models have difficulty learning the additional tag types; The precision for H- and D-tags is consistently lower than the precision for B-tags. In fact, on the PsyTAR data which contains only few overlapping entities (1.7%), the H-tag was never predicted. It seems that FuzzyBIO makes the task easier in two ways, namely by standardizing entities into continuous sequences that always start with a B-tag and by excluding rare tags such as the H-tag. Standardizing entities makes it easier for the model to learn the underlying rules and excluding rare tags removes a goal for which there are only few examples available.

7 Conclusion

As recommendation for future NER tasks, we expect FuzzyBIO to especially benefit NER for difficult tasks with a fair amount of discontinuous entities. However, since the conversion from BIOHD to FuzzyBIO is straightforward and deterministic, we recommend to experimentally compare which of the two representations is more effective for any data set that includes discontinuous entities.

Acknowledgments

We would like to thank the SIDN fonds for funding this research and our reviewers for their valuable feedback.

References

Lorna Hazell and Saad A W Shakir. 2006. Under-Reporting of Adverse Drug Reactions A Systematic Review. *Drug Safety*, 29(5):385–396.

Sarvnaz Karimi, Alejandro Metke-Jimenez, Madonna Kemp, and Chen Wang. 2015. Cadec: A corpus of adverse drug event annotations. *Journal of Biomedical Informatics*, 55:73–81.

Jean Baptiste Lamy, Alain Venot, and Catherine Duclos. 2015. PyMedTermino: An open-source generic API for advanced terminology services. *Studies in Health Technology and Informatics*, 210:924–928.

Jérémy Lardon, Redhouane Abdellaoui, Florelle Bellet, Hadyl Asfari, Julien Souvignet, Nathalie Texier, Marie Christine Jaulent, Marie Noëlle Beyens, Anita Burgun, and Cédric Bousquet. 2015. Adverse drug reaction identification and extraction in social media: A scoping review. *Journal of Medical Internet Research*, 17(7):1–16.

Arjun Magge, Elena Tutubalina, Zulfat Miftahutdinov, Ilseyar Alimova, Anne Dirkson, Suzan Verberne, Davy Weissenbacher, and Graciela Gonzalez-Hernandez. 2020. Deepademiner: A deep learning pharmacovigilance pipeline for extraction and normalization of adverse drug effect mentions on twitter. *medRxiv*.

Arun S. Maiya. 2020. ktrain: A low-code library for augmented machine learning. *arXiv*, arXiv:2004.10703 [cs.LG].

Alejandro Metke-Jimenez and Sarvnaz Karimi. 2015. Concept Extraction to Identify Adverse Drug Reactions in Medical Forums: A Comparison of Algorithms. *ArXiv*.

Natalya F. Noy, Nicholas Griffith, and Mark A. Musen. 2008. Collecting community-based mappings in an ontology repository. In *Proceedings of the 7th International Conference on The Semantic Web*, ISWC '08, page 371–386, Berlin, Heidelberg. Springer-Verlag.

Sameer Pradhan, Noémie Elhadad, Wendy Chapman, Suresh Manandhar, and Guergana Savova. 2014. SemEval-2014 Task 7: Analysis of Clinical Text. In *Proceedings ofthe 8th International Workshop on Semantic Evaluation (SemEval 2014)*, SemEval, pages 54–62.

Victor Sanh, Lysandre Debut, Julien Chaumond, and Thomas Wolf. 2019. DistilBERT, a distilled version of BERT: smaller, faster, cheaper and lighter. In *5th Workshop on Energy Efficient Machine Learning and Cognitive Computing - NeurIPS 2019*.

Abeed Sarker, Rachel Ginn, Azadeh Nikfarjam, Karen O'Connor, Karen Smith, Swetha Jayaraman, Tejaswi Upadhaya, and Graciela Gonzalez. 2015. Utilizing social media data for pharmacovigilance: A review. *Journal of Biomedical Informatics*, 54:202–212.

Jingbo Shang, Liyuan Liu, Xiaotao Gu, Xiang Ren, Teng Ren, and Jiawei Han. 2020. Learning named entity tagger using domain-specific dictionary. *Proceedings of the 2018 Conference on Empirical Methods in Natural Language Processing, EMNLP 2018*, pages 2054–2064.

Premnath Shenoy and Anand Harugeri. 2015. Elderly patients' participation in clinical trials. *Perspectives in Clinical Research*, 6(4):184.

Yuqi Si, Jingqi Wang, Hua Xu, and Kirk Roberts. 2019. Enhancing clinical concept extraction with contextual embeddings. *Journal of the American Medical Informatics Association*, 26(11):1297–1304.

Leslie N. Smith. 2018. A disciplined approach to neural network hyper-parameters: Part 1 – learning rate, batch size, momentum, and weight decay. *ArXiv*.

Jennifer D ' Souza and Vincent Ng. 2015. Sieve-Based Entity Linking for the Biomedical Domain. In *53rd Annual Meeting ofthe Association for Computational Linguistics and the 7th International Joint Conference on Natural Language Processing*, pages 297–302.

Mujeen Sung, Hwisang Jeon, Jinhyuk Lee, and Jaewoo Kang. 2020. Biomedical Entity Representations with Synonym Marginalization. In *Proceedings of the 58th Annual Meeting of the Association for Computational Linguistics*, pages 3641–3650.

Buzhou Tang, Qingcai Chen, Xiaolong Wang, Yonghui Wu, Yaoyun Zhang Phd, Min Jiang, Jingqi Wang, and Hua Xu. 2015. Recognizing Disjoint Clinical Concepts in Clinical Text Using Machine Learning-based Methods. In *AMIA Annu Symp Proc.*, pages 1184–1193.

Buzhou Tang, Jianglu Hu, Xiaolong Wang, and Qingcai Chen. 2018. Recognizing Continuous and Discontinuous Adverse Drug Reaction Mentions from Social Media Using LSTM-CRF. *Wireless Communications and Mobile Computing*, 2018.

Elena Tutubalina, Artur Kadurin, and Zulfat Miftahutdinov. 2020. Fair Evaluation in Concept Normalization : a Large-scale Comparative Analysis for BERT-based Models. In *COLING 2020*.

Özlem Uzuner, Brett R. South, Shuying Shen, and Scott L. DuVall. 2011. 2010 i2b2/VA challenge on concepts, assertions, and relations in clinical text. *Journal of the American Medical Informatics Association*, 18(5):552–556.

Davy Weissenbacher, Abeed Sarker, Arjun Magge, Ashlynn Daughton, Karen O'Connor, Michael J. Paul, and Graciela Gonzalez-Hernandez. 2019. Overview of the fourth social media mining for health (SMM4H) shared tasks at ACL 2019. In *Proceedings of the Fourth Social Media Mining for Health Applications (#SMM4H) Workshop & Shared Task*, pages 21–30, Florence, Italy. Association for Computational Linguistics.

Thomas Wolf, Lysandre Debut, Victor Sanh, Julien Chaumond, Clement Delangue, Anthony Moi, Pierric Cistac, Tim Rault, Rémi Louf, Morgan Funtowicz, and Jamie Brew. 2019. Transformers: State-of-the-art Natural Language Processing. *ArXiv*.

World Health Organisation. 2006. The Safety of Medicines in Public Health Programmes: Pharmacovigilance an essential tool. Technical report, World Health Organisation.

Yaoyun Zhang, Jingqi Wang, Buzhou Tang, Yonghui Wu, Min Jiang, Yukun Chen, and Hua Xu. 2014. UTH_CCB: A report for SemEval 2014 – Task 7 Analysis of Clinical Text. In *Proceedings ofthe 8th International Workshop on Semantic Evaluation (SemEval 2014)*, pages 802–806.

Maryam Zolnoori, Kin Wah Fung, Timothy B Patrick, Paul Fontelo, Hadi Kharrazi, Anthony Faiola, Nilay D Shah, Yi Shuan Shirley Wu, Christina E Eldredge, Jake Luo, Mike Conway, Jiaxi Zhu, Soo Kyung Park, Kelly Xu, and Hamideh Moayyed. 2019. The PsyTAR dataset: From patients generated narratives to a corpus of adverse drug events and effectiveness of psychiatric medications. *Data in brief*, 24.

Cluster Analysis of Online Mental Health Discourse using Topic-Infused Deep Contextualized Representations

Atharva Kulkarni[1], **Amey Hengle**[1], **Pradnya Kulkarni, Ph.D**[2], and **Manisha Marathe**[3]

[1, 3]Department of Computer Engineering, PVG's College of Engineering and Technology, affiliated to Savitribai Phule Pune University, India.
[2]Department of Psychology, Sir Parashurambhau College, Pune, India.
[1]{atharva.j.kulkarni1998, ameyhengle22}@gmail.com
[2]pradnya.kulkarni@spcollegepune.ac.in
[3]mvm_comp@pvgcoet.ac.in

Abstract

With mental health as a problem domain in NLP, the bulk of contemporary literature revolves around building better mental illness prediction models. The research focusing on the identification of discussion clusters in online mental health communities has been relatively limited. Moreover, as the underlying methodologies used in these studies mainly conform to the traditional machine learning models and statistical methods, the scope for introducing contextualized word representations for topic and theme extraction from online mental health communities remains open. Thus, in this research, we propose topic-infused deep contextualized representations, a novel data representation technique that uses autoencoders to combine deep contextual embeddings with topical information, generating robust representations for text clustering. Investigating the Reddit discourse on Post-Traumatic Stress Disorder (PTSD) and Complex Post-Traumatic Stress Disorder (C-PTSD), we elicit the thematic clusters representing the latent topics and themes discussed in the r/ptsd and r/CPTSD subreddits. Furthermore, we also present a qualitative analysis and characterization of each cluster, unraveling the prevalent discourse themes.

1 Introduction

Due to their ubiquitous nature, online health communities and social media platforms have emerged as a conducive means of information exchange and social support, especially for people with stigmatized concerns such as mental health. Consequently, these platforms provide a rich ecosystem for mental health clinicians, researchers, and practitioners to analyze the cornucopia of user-generated content and study the underlying mechanisms of different mental health conditions. With the rapid headways in artificial intelligence and computational linguistics, an increasingly large number of researchers have leveraged social media content to study various mental health illnesses and psychiatric conditions. While most of the research has focused on using the traditional machine learning models and statistical methods for predicting mental illness from social media posts, the studies addressing the discourse analysis (De Choudhury and De, 2014; Silveira Fraga et al., 2018; Loveys et al., 2018) and identification of clusters in online mental health communities (Park et al., 2018) has been relatively modest. Even with the recent surge of complex attention-based deep learning models, a large chunk of the research regarding mental health issues has focused on building better predictive systems (Benton et al., 2017; Kirinde Gamaarachchige and Inkpen, 2019; Jiang et al., 2020; Sekulic and Strube, 2019) with less emphasis on using these models for mental health related corpus analysis or information extraction.

With mental health already coalesced as an appreciable public health burden, research to investigate the discourse clusters prevalent on social media is of paramount importance. Mining information from the emergent clusters provides a lens over the dominant themes of discussion, the discourse anatomy, and the dialogue structure in the online forums while also helping to comprehend the general public engagement, sentiment, ideas, and views regarding mental health. Successful research in this direction can potentially foster identification of high-risk groups, enhanced mental health patient education programs, better diagnostic and therapeutic theory building, as well as an improved understanding of the underlying design of the online mental health communities (Park et al., 2018).

Post-traumatic stress disorder (PTSD) is a mental disorder resulting from traumatic experiences that leads to reliving the trauma, avoidance of certain situations, and hyper-vigilance. Similar to PTSD, complex post-traumatic stress disorder (C-

Proceedings of the 12th International Workshop on Health Text Mining and Information Analysis, pages 83–93
April 19, 2021.

PTSD) is a condition that formulates the reaction resulted from the trauma, such as uncontrollable emotions, dissociation, negative self-perception, anger, mistrust, and interpersonal difficulties. Thus, in this research, we examine the online discourse on Reddit, focussing on PTSD and C-PTSD as use cases to elicit different thematic clusters present in them.

Prior research has shown that the addition of topic information to pre-trained contextualized representations yields performance improvement for semantic textual similarity (Peinelt et al., 2020). While Peinelt et al. (2020) integrated the two representations by simple concatenation, a better methodology to integrate representations from different embedding spaces is argued by Bollegala and Bao (2018). The meta-embeddings proposed by Bollegala and Bao (2018) are learned as the intermediate representations generated by various autoencoder variants. Thus, building on these two findings, we propose topic-infused deep contextualized representations, a novel data representation technique that uses a concatenated denoise autoencoder to combine deep contextual embeddings with topic information for generating robust document representations. Our methodology spawns document representations that subsume the topic information from Latent Dirichlet Allocation (Blei et al., 2003) with the contextual embeddings generated by pre-trained RoBERTa model(Liu et al., 2019). We further demonstrate that the proposed methodology achieves improvement for text clustering against the contextual embeddings generated by the pre-trained RoBERTa model.

In the light of the above discussion, our research makes the following contributions:

- We extend the methodology of word meta-embeddings to document meta-embeddings by proposing topic-infused deep contextualized representations, a data representation technique that uses a concatenated denoise autoencoder to combine deep contextual embeddings with topical information for generating robust representations for text clustering.

- We carry out a qualitative analysis and characterization of each cluster from a clinical psychology perspective.

Attribute	r/ptsd	r/CPTSD	Total
Total posts	28,133	69,600	97,733
Filtered posts	4,511	20,419	24,930
Average post length	213.52	240.36	235.51
Average comments per post	14.17	15.58	15.33
Average net upvotes per post	39.94	60.34	56.65

Table 1: Dataset statistics.

2 Related Work

Traditionally, the research in topic mining and theme extraction from online mental health communities has been focused on the use of probabilistic generative models like the Latent Dirichlet Allocation (Blei et al., 2003) and clustering techniques such as the k-means (Schütze et al., 2008). Carron-Arthur et al. (2016) employed LDA to extract topics from the internet support group BlueBoard. Results showed that users engaged in discussions with a greater topical focus on experiential knowledge, disclosure, and informational support, a pattern resembling the clinical symptom-focused approach to recovery. In their study, Dao et al. (2017) used the Hierarchical Dirichlet Process (HDP) algorithm to infer latent topics from blog posts of the Live-Journal (LJ) blogging site. The authors applied the non-parametric affinity propagation algorithm to find clusters within the online communities. Toulis and Golab (2017) compared the recurring themes encountered in private journals with the ones found in the online communities of Reddit and found significant similarities in the topics discussed across both the forums. Park et al. (2018) provide an exhaustive analysis of the thematic overlap, similarity, and difference in online mental health communities of r/depression, r/anxiety, and r/ptsd. Their results show a considerable overlap of themes between the mental health groups, attesting that people engaging in such forums face common problems and comorbidity symptoms.

Since their introduction, transformer-based language models such as BERT (Devlin et al., 2019) have led to impressive performance gains across multiple NLP tasks. Recent works show that these contextualized representations can also support type-level clusters, and hence, can be effectively used for modeling topics (Sia et al., 2020). The recent work by Thompson and Mimno (2020) demonstrates that running simple clustering algorithms

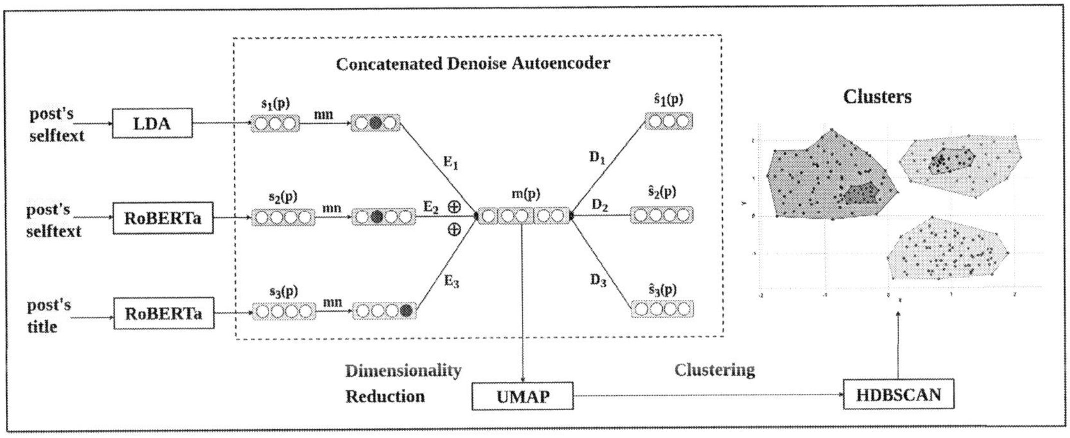

Figure 1: Architecture of the proposed system. Here, *mn* refers to masking noise applied to the autoencoder inputs.

like k-means on contextualized word representations result in word clusters that share similar properties to the ones generated by LDA. An interesting approach is followed by Peinelt et al. (2020), where the authors combine the topic models of LDA with contextualized word representations of BERT for the task of semantic similarity detection. Results depict that adding topical information improves performance, especially for examples with domain-specific terms.

3 Dataset Description

For this experiment, we selected the r/ptsd[1] and r/CPTSD [2] subreddits which have 54,000 and 97,000 active users, respectively. Using the Pushift API [3], we crawled all the available posts from these subreddits between 1st August 2015 and 31st July 2020.

To ensure that each selected post has community approval, we selected posts that have a minimum of 10 net upvotes. We further filter our dataset by eliminating posts with less than 75% English content [4], posts with less than five words, as well as posts with [DELETED], [UPDATED], and [REMOVED] entries. We employed standard text cleaning and normalization techniques for preprocessing the posts, including removing special characters, accented words, wordplays, URLs, replacing acronyms with full forms, and expanding contractions[5]. This resulted in a comprehensive dataset of 24,930 posts.

[1] https://www.reddit.com/r/ptsd/
[2] https://www.reddit.com/r/CPTSD/
[3] https://pushshift.io/
[4] https://pypi.org/project/pycld2/0.24/
[5] https://pypi.org/project/pycontractions/

The dataset statistics are provided in Table 1.

4 Proposed Methodology

The proposed system consists of two key components: a robust data representation methodology and an efficient clustering algorithm. Figure 1 depicts the model architecture. Each component is elucidated in detail as follows:

4.1 Topic-infused Deep Contextualized Representations

In this section, we posit the methodology to generate the topic-infused contextualized representations, using a multi-input concatenated denoise autoencoder. The proposed autoencoder architecture has three inputs namely: the document topic distribution of the post's selftext [6], contextualized document embedding of the post's selftext, and the contextualized document embedding of the post's title. Let S_1, S_2, and S_3 denote the corresponding three input embedding spaces of dimensions d_1, d_2, and d_3, respectively. Let N be the total number of posts. For each post $p \in N$, the three document embeddings are given by $s_1(p) \in \mathbb{R}^{d_1}$, $s_2(p) \in \mathbb{R}^{d_2}$, and $s_3(p) \in \mathbb{R}^{d_3}$. The autoencoder model consists of three encoders E_1, E_2, and E_3 which encode the source embeddings to a common meta-embedding space M of dimensionality d_m. Each encoder independently performs dimensionality reduction and non-linear transformations on the respective embeddings, thus, learning to retain essential information from each source embedding. Dimensionalities of the encoded input embeddings are denoted respectively, by d_1', d_2',

[6] selftext attribute refers to the main body of the post.

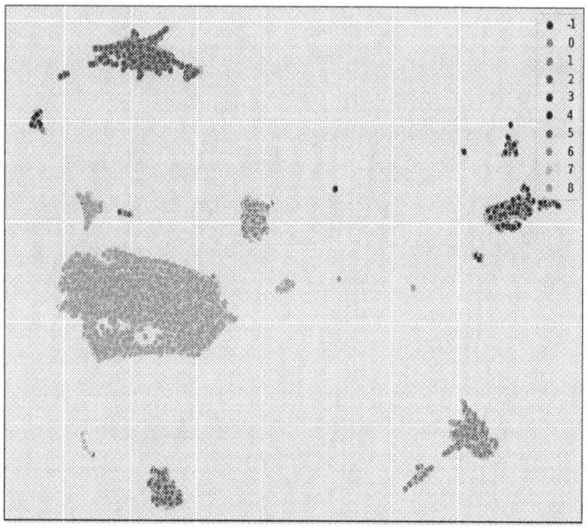

Figure 2: 2D embdding space of the topic-infused deep contextualized representations (-1 represents the unassigned datapoints).

and d'_3. The concatenation of each of the encoded input source embeddings results in the document meta-embedding $m(p)$, as given by Equation 1.

$$m(p) = E_1(s_1(p)) \oplus E_2(s_2(p)) \oplus E_3(s_3(p)) \quad (1)$$

Therefore, the dimensionality of document meta-embedding space M is computed as the sum of dimensionalities of encoded source embeddings. It is given by Equation 2.

$$d_m = d'_1 + d'_2 + d'_3 \quad (2)$$

The three decoders D_1, D_2, and D_3, try to reconstruct the individual source embeddings from the document meta-embeddings, thereby implicitly utilizing the common and the complementary information present in the source embeddings. Equations 3, 4, and 5 represent the reconstructed versions of the source embeddings, given by $\hat{s}_1(p)$, $\hat{s}_2(p)$, and $\hat{s}_3(p)$.

$$\hat{s}_1(p) = D_1(m(p)) \quad (3)$$

$$\hat{s}_2(p) = D_2(m(p)) \quad (4)$$

$$\hat{s}_3(p) = D_3(m(p)) \quad (5)$$

The objective loss $\mathcal{L}$ is calculated as the sum of the reconstruction loss for each of the three input embeddings. It is formulated in Equation 6.

$$\mathcal{L} = \sum_{p \in N} (||\hat{s}_1(p) - s_1(p)||^2 + ||\hat{s}_2(p) - s_2(p)||^2 + ||\hat{s}_3(p) - s_3(p)||^2) \quad (6)$$

Cluster	Original cluster size	Exemplar cluster size
Cluster 1	360	300
Cluster 2	1756	300
Cluster 3	2829	600
Cluster 4	361	300
Cluster 5	1971	300
Cluster 6	1495	300
Cluster 7	1348	300
Cluster 8	417	300
Cluster 9	14065	2100
Unassigned	46	-

Table 2: Cluster and respective exemplar sizes.

Thus, the proposed system jointly learns E_1, E_2, E_3, and D_1, D_2, D_3, such that the loss given by Equation 6 is minimized.

4.2 Dimensionality Reduction and Clustering

In this study, we make use of HDBSCAN (Hierarchical Density-Based Spatial Clustering of Applications with Noise), a hierarchical density-based unsupervised clustering technique to congregate semantically-similar posts together in clusters (McInnes et al., 2017). Unlike partition-based clustering algorithms, HDBSCAN can find varying density clusters and is more robust to parameter selection. Moreover, HDBSCAN does not force the data points to belong to any particular cluster, making it suitable for handling outliers and noisy data.

As HDBSCAN uses relative-distance measures for clustering, it often suffers from the curse of dimensionality (McInnes et al., 2017). As the dimension of the topic-infused contextualized embeddings is quite high (d_m = 768), we employ UMAP (Uniform Manifold Approximation Projection), a technique for general non-linear dimensionality reduction (McInnes et al., 2018). UMAP is preferred over other dimensionality reduction algorithms as it keeps a significant portion of the high-dimensional local structure in lower dimensionality, thus, causing minimal information loss.

5 Experimental Setup

The document topic distributions are generated using the LDA mallet python version [7]. As the topics of interest centers around entities that are mostly

[7] `https://radimrehurek.com/gensim/models/wrappers/ldamallet.html`

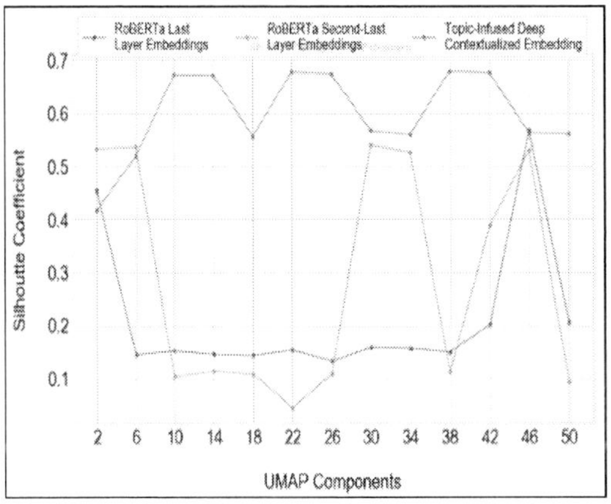

Figure 3: Silhoutte Coefficient (SC) Comparison

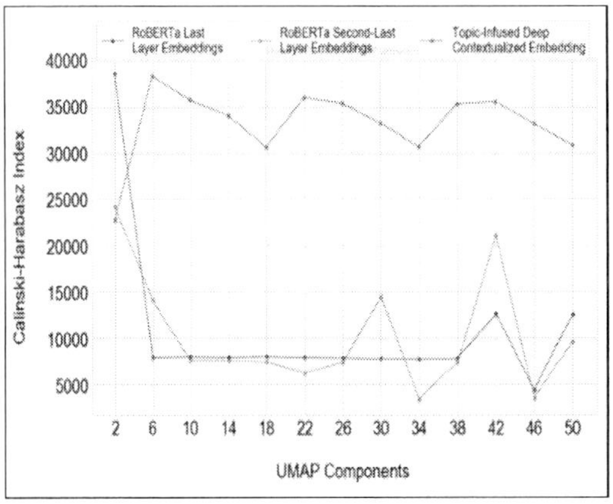

Figure 4: Calinski Harabasz Index (CHI) Comparison

nouns, we follow a nouns-only approach as employed by Martin and Johnson (2015) for topic modeling. The number of topics is empirically chosen as 10 since it displayed the best topic coherence score (c_v) of 0.49. The rest of the hyperparameters for LDA are kept at default. The 768-dimensional contextualized document embeddings are generated as the average of all the embeddings for each word in the document, extracted from the second last layer of the pre-trained RoBERTa-base model (Liu et al., 2019). Thus, the three input d_1, d_2, and d_3 are of dimensions 10, 768, and 768, respectively.

In our experiments, each autoencoder is implemented as a single hidden layer neural network. The dimensions d'_1, d'_2, and d'_3 are chosen as 10, 379, and 379, respectively. The hidden dimensions are chosen as such so that the topic-infused deep contextualized representations are of dimensions 768, making them comparable with that of RoBERTa. Masking noise of 10 percent is applied to the source embeddings before encoding (Vincent et al., 2010). Leaky rectified linear (Leaky ReLU) (Maas et al., 2013) activation is applied to each layer with the default parameter setting. The model is trained end-to-end for 200 epochs, with the Adam optimizer (Kingma and Ba, 2015), a learning rate of 0.001, and a mini-batch size of 128 for minimizing the objective loss. The learning rate is reduced by a factor of 0.1 if validation loss does not decline after 10 successive epochs. The model with the best loss is used for prediction.

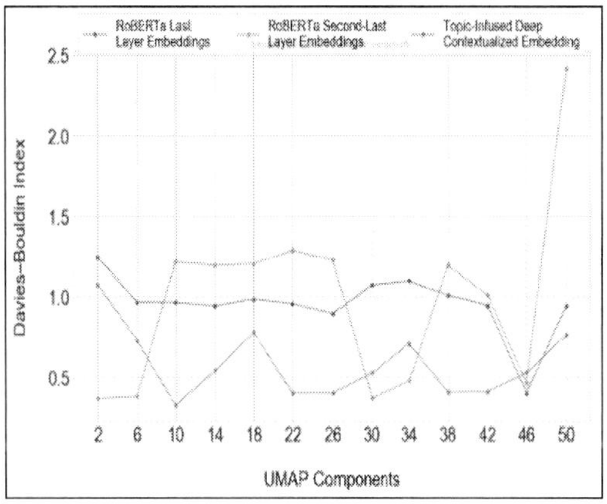

Figure 5: Davies Bouldin Index (DBI) Comparison

6 Clustering Performance Evaluation

In order to assess the clustering performance, we make use of three measures, namely the Silhouette Coefficient (SC) (Rousseeuw, 1987), the Calinski-Harabasz Index (CHI) (Caliński and Harabasz, 1974), and the Davies-Bouldin Index (DBI) (Davies and Bouldin, 1979). Higher SC, and CHI, and lower DBI scores indicate a better separation of the clusters and tighter integration inside the clusters. We compare the clustering performance of three embedding sets, namely: the RoBERTa last layer embeddings, the RoBERTa second last layer embeddings, and our proposed topic-infused deep contextualized embeddings. Figures 3, 4, and 5 depict the performance comparison of the corresponding three embeddings with respect to the three metrics mentioned above. The y-axis represents the three corresponding metrics, whereas the x-axis represents the UMAP component variations. The hyperparameter values of the number of neighbours and minimum distance were chosen as 30 and 0.0, respectively, while cosine similarity was used as the metric for computing the distance in the ambient space of the input data. The random state is seeded to an integer value (42). For HDB-

SCAN, the minimum cluster size was selected as 300, with the other parameters set to the default values (McInnes et al., 2017). From the comparative analysis, it is evident that the proposed topic-infused deep contextualized representations result in an improved clustering performance across all three metrics. For our study, we choose UMAP with components 10, as it empirically gives consistent performance across all three metrics. The 2D embedding space after clustering is shown in Figure 2.

To reduce the cluster size, we extract the most representative posts or exemplars of each cluster [8]. In technical terms, exemplars are the data points that lie at the heart of a cluster, around which the ultimate cluster forms. Table 2 provides the statistics of the clusters and their respective exemplar sizes. To find the key-words for each cluster, we employ a class-based term-frequency inverse-document-frequency (c-TF-IDF)[9] method on each of the obtained exemplars. Unlike the traditional TF-IDF, which considers each document of a corpus, c-TF-IDF is a class-based method that treats all the documents belonging to a particular class as a single document. This enables us to find only the latent topics most representative of a particular cluster and penalize the frequent words across the clusters. For each cluster, the c-TF-IDF score is calculated using the Equation 7, where each word t is extracted for each class i, and the number of documents m is divided by the total frequency of the word t across all classes n.

$$c-TF-IDF_i = \frac{t_i}{w_i} * \log \frac{m}{\sum_{j=1}^{n} t_j} \quad (7)$$

The coherence scores for the topics generated by LDA and the clusters generated by the above-mentioned embeddings are reported in Table 3. The top 15 nouns with the highest c-TF-IDF scores for each cluster are used to evaluate the topic coherence. The overall coherence of the topics generated is evaluated using extrinsic as well as intrinsic measures. As the intrinsic topic coherence is computed using word co-occurrences in documents from the corpus (Mimno et al., 2011), it is natural that LDA reports the highest intrinsic coherence (c_v), as it is based on word-occurrences statistics. Moreover, as argued by Stevens et al. (2012), a good intrinsic coherence score does not necessarily guarantee that the generated topics make semantic sense or that they are interpretable by humans. *Normalized pointwise mutual information* (NPMI) (Bouma, 2009), on the other hand, has been shown to correlated with human judgement (Lau et al., 2014). Thus, in this study, we use NPMI as an extrinsic measure. Experiments by Sia et al. (2020) suggest that clustering contextual embeddings can result in topics with better NPMI compared to LDA. As evident from the results reported in Table 3, our results corroborate these findings, as our proposed methodology exhibits the best NPMI score. Figure 6 depicts the wordclouds of the top 50 nouns with the highest c-TF-IDF scores for each cluster.

Embeddings	**Extrinsic Measure**	**Intrinsic Measure**
RoBERTa last layer	-0.0678	0.2650
RoBERTa 2nd last layer	-0.0500	0.2702
Topic-infused deep contextualized representations	**-0.0255**	0.3186
LDA	-0.2994	**0.4972**

Table 3: Topic coherence results.

7 Thematic Analysis of the Clusters

For a comprehensive qualitative evaluation, we select the top 100 posts from each clusters which exhibit the highest Jaccard Index scores [10] with that of the top 15 nouns with the highest c-TF-IDF scores for each cluster. The qualitative analysis is carried out by the subject expert in clinical psychology to draw out the major discussion themes. Total 7 themes were extracted from the 9 clusters generated. They are as follows:

Abuse stories: Cluster 3 reveals self-disclosure stories of abuse that are physical, emotional, or sexual in nature. Abusers are mainly parents, siblings, close family members, co-workers, and known people. Narrations prominently highlight dysfunctional family dynamics during the victim's childhood as well as the issues related to current family interactions. Family dynamics are characterized by alcoholic/ substance abusive parents and abusers with pathological personality traits. Posts indicate traumatic experiences, the effect of these experiences, and how they act as triggers. The victim's exposure to the trauma was long and frequent.

[8] https://hdbscan.readthedocs.io/en/latest/soft_clustering_explanation.html

[9] https://github.com/MaartenGr/cTFIDF

[10] https://en.wikipedia.org/wiki/Jaccard_index

Figure 6: Wordclouds of top 50 c-TF-IDF nouns for the 9 clusters.

> *"My **mother** was very **controlling, scary, abusive and manipulative**..."*
> *"I've been coming to terms over the past 6 months or so with the fact that my **family** caused me a significant amount of **trauma growing up**..."*
> *"For as long as I can remember, my **father** was **physically and emotionally abusive**..."*

Flashbacks: The primary themes of cluster 6 include flashbacks of traumatic incidences and the emotional disturbances associated with them. Feelings of anxiety, panic, nightmares, fear, impulsivity, and anger are prevalent throughout the cluster. Narrations reveal the way flashbacks are impeding current life issues. Examples include an aversion to touch and sexual experiences, difficulties in romantic relationships, social relationships, daily chores, and work as well as triggering health-related symptoms. Working through flashbacks, ways of dealing with it, and help to overcome it are also shared.

> *"...struggling with **flashbacks, nightmares, and intrusive thoughts** about **trauma**. can anyone give me reassurance that i'll get through this?..."*

> *"...I can stay that way for weeks while I slowly process the **old trauma** that comes up. I'm in that state right now - just **overwhelmed** by **intense emotional flashbacks**..."*

Advice seeking: People have vividly expressed their feelings and are seeking advice about various issues related to PTSD, according to the posts in cluster 8. Main themes revolve around advice related to job functioning impacted by PTSD symptoms, social isolation, dependency issues, management of difficult emotions, exhaustion, and frustration. Posts in this cluster are brief and direct, seeking advice, help, and support from the Reddit community.

> *"...I would appreciate any **tips or advice**. **How do you** deal with emotional isolation?"*
> *"When I'm more myself, I feel like I'm too cynical and rude. **Anyone else** have experience with this? **How do you** become more authentic? "*

Therapy and therapist experiences: Cluster 1 and 4 mainly focuses on therapy and therapist experiences. Posts highlight the process of therapy, experiences with therapists, challenges faced in the therapeutic process, insights from therapies, transference experiences, and seeking help for such issues. Other findings include posts that seek therapeutic methods other than visiting therapists. Surprisingly, a large chunk of the posts portrays a negative connotation about therapy experiences or therapy.

> *"I've been **traumatized by therapists** in my childhood that relayed info back to my abusive parents and gave them more ammo to hurt me with..."*
>
> *"The only **therapist** who is willing to treat me is **unreliable, too far away, and frankly unsympathetic and unqualified**..."*

Difficulty in emotional regulation: Posts in cluster 5 primarily focus on different emotions associated with trauma. Posts suggest difficulty in controlling emotions, emotional neglect, panic, anxiety, feeling of emptiness, hopelessness, emotional distancing due to trauma, difficulty understanding and expressing emotions, and ambivalent emotions towards abusers. Other findings include the progress in trauma-related emotions. Most of the posts are help-seeking in nature, for validating their emotions.

> *"DAE **struggle** with identifying or verbally explaining your **emotions**?"*
>
> *"I have recently realized that I have the **hardest time identifying my emotions**..."*
>
> *"DAE struggle with **guilt** around cutting off **toxic people**?"*
>
> *"How did you overcome chronic **emotional numbness**? Inability and lack of desire to develop a connection with others (platonic or romantic)?"*
>
> *"DAE feel like they've **built a wall** between themselves and their **feelings**?"*

Working through traumatic abuse: Post in cluster 2 narrates retrospective abusive, traumatic experiences and their psychological after-effects. The majority of the posts indicate signs of shame, dissociation, social withdrawal, anxiety, self-downing, and victimization. Confrontation, triumph while struggling with vulnerabilities, hope, and new insights about oneself are other findings.

> *"Mainly the reasons I **self-isolate, distrust/avoid** intimacy, and deprive myself of certain pleasures is due to my PTSD symptoms because of all the **abuse** I endured..."*
>
> *"The lifelong **shame and fear**, all the problems causing me to not succeed in life or have any healthy relationships..."*
>
> *"I haven't lived a day in my adult life. It's all been a decades-long **dissociation**."*

Abuse issues and general PTSD: Clusters 7 and 9 mostly contains posts about abuse stories and PTSD in general. The major themes are related to the parent-child issues, insecure parenting styles, emotional and physical abuse, the effect of childhood trauma, loss and grieving for not having healthy parental relations, the impact of insecure parenting, disclosure of trauma incidences, and sharing recovery tips. Following are a few illustrations:

> *".. I **hate** the days where you long for the **childhood** that you never had and **support you never received**. For the belonging and "home" that you never had ..."*
>
> *"... My last therapy session had me divulging my **experiences as a young child** and what life was like for me back then..."*

8 Discussions

Cluster structure is in accordance with the formal diagnostic criteria of PTSD and C-PTSD. Abuse stories reflected in cluster 1 corresponds to the *criterion-A* of PTSD in DSM-5 (Association et al., 2013), that states the experience of trauma. Out of the six symptom cluster proffered by ICD-11[11] for CPTSD (re-experiencing, avoidance, hypervigilance, emotional dysregulation, interpersonal difficulties, and negative self-concept), cluster 6 adequately corresponds to re-experiencing of flashbacks, while cluster 5 is consistent with emotional dysregulation. Parent-child relationships portrayed in cluster 2 corresponds to etiological factors. Dysfunctional parent-child relation in PTSD has been widely confirmed in the clinical psychology literature (Cockram et al., 2010; Cross et al., 2018; van Ee et al., 2016). On the other hand, clusters 5, 6, and 7 characterizing therapy experience, working

[11] `https://icd.who.int/en`

through trauma, and advice, respectively, attribute to the treatments and interventions aspect of PTSD. In conclusion, it can be said that these clusters reveal salient features of PTSD.

Although the clusters more or less convey their discrete independent themes, there exist many common topics running throughout the clusters. The recurrent expressions observed throughout the clusters pertain to **personal and social life** *(family, parents, friends, person, relationships)*, **daily grind** *(work, home, job, place)*, and **temporal indicators** *(day, time, year, month, today, week)*. Furthermore, words encompassing **cognition** *(thoughts, think, feel, feeling, lot, hard)*, **emotional and affect expressions** *(happy, love, good, anxiety, kind, bad, hate)*, and **inhibition expressions** *(avoid, deny, escape)* are perpetual throughout the posts.

9 Limitations

Our work has its own drawbacks and limitations. The problem statement at hand is of fine-grained clustering rather than coarse-grained clustering, thus, making it difficult to draw out interpretable topics of discussion. This is attested by the low NPMI scores in Table 3. Furthermore, the posts' lengthy nature (refer Table 1) makes it difficult for models like RoBERTa to capture maximum contextual information (Liu et al., 2019). Many psychiatric disorders are found to be co-morbid with PTSD (Brady et al., 2000). The research by Park and Conway (2018) indicates that people facing mental health issues often find it difficult to regulate their ideas and views. Their research further entails that posts from online mental health communities are more difficult to read and portray less lexical diversity. These factors further hinder the elicitation of interpretable topics from the corpus. Also, not all clusters exhibit independent themes of discussions and some themes are a combination of multiple clusters.

10 Conclusion and Future Work

In this research, we introduce topic-infused deep contextualized representations, a robust data representation methodology that successfully integrates topical information with contextualized embeddings. We collect and analyze large-scale data pertaining to the online discourse on Reddit, centered around PTSD and CPTSD, and perform density-based clustering to draw out the prominent clusters and analyze the themes of discussion present in them. Despite the perpetual semantic and thematic similarities amongst the posts in the corpus, our methodology, to some extent, is able to draw out the underlying, fine-grained, latent clusters. The qualitative analysis of each cluster revealed the characteristic themes and salient features of PTSD and CPTSD, consistent with the clinical psychology literature. Through the lens of social media, this study delineates a deeper understanding of PTSD and C-PTSD, fostering further research in early detection of mental illnesses, identification of high-risk groups, enhanced mental health patient education programs, better diagnostic and therapeutic theory building, as well as an improved understanding of the underlying design of the online mental health communities.

The future work of this research can take multiple directions. From an NLP standpoint, the robustness of the proposed methodology should be examined by testing them on various other benchmark NLP tasks such as semantic textual similarity, word analogy, and text classification, to name a few. Other variants of autoencoders and objective losses could be employed to facilitate tighter integration of topical information with the contextualized embeddings. From the mental health and clinical psychology perspective, such research can be easily extended to other online mental health communities to draw useful insights.

References

American Psychiatric Association et al. 2013. *Diagnostic and statistical manual of mental disorders (DSM-5®)*. American Psychiatric Pub.

Adrian Benton, Margaret Mitchell, and Dirk Hovy. 2017. Multitask learning for mental health conditions with limited social media data. In *Proceedings of the 15th Conference of the European Chapter of the Association for Computational Linguistics: Volume 1, Long Papers*, pages 152–162, Valencia, Spain. Association for Computational Linguistics.

David M Blei, Andrew Y Ng, and Michael I Jordan. 2003. Latent dirichlet allocation. *Journal of machine Learning research*, 3(Jan):993–1022.

Danushka Bollegala and Cong Bao. 2018. Learning word meta-embeddings by autoencoding. In *Proceedings of the 27th International Conference on Computational Linguistics*, pages 1650–1661, Santa Fe, New Mexico, USA. Association for Computational Linguistics.

Gerlof Bouma. 2009. Normalized (pointwise) mutual

information in collocation extraction. *Proceedings of GSCL*, pages 31–40.

Kathleen T Brady, Therese K Killeen, Tim Brewerton, and Sylvia Lucerini. 2000. Comorbidity of psychiatric disorders and posttraumatic stress disorder. *The Journal of clinical psychiatry*, 61(suppl 7):22–32.

T. Caliński and J Harabasz. 1974. A dendrite method for cluster analysis. *Communications in Statistics*, 3(1):1–27.

Bradley Carron-Arthur, Julia Reynolds, Kylie Bennett, Anthony Bennett, and Kathleen M Griffiths. 2016. What's all the talk about? topic modelling in a mental health internet support group. *BMC psychiatry*, 16(1):367.

David M Cockram, Peter D Drummond, and Christopher W Lee. 2010. Role and treatment of early maladaptive schemas in vietnam veterans with ptsd. *Clinical Psychology & Psychotherapy: An International Journal of Theory & Practice*, 17(3):165–182.

Dorthie Cross, L Alexander Vance, Ye Ji Kim, Andrew L Ruchard, Nathan Fox, Tanja Jovanovic, and Bekh Bradley. 2018. Trauma exposure, ptsd, and parenting in a community sample of low-income, predominantly african american mothers and children. *Psychological trauma: theory, research, practice, and policy*, 10(3):327.

Bo Dao, Thin Nguyen, Svetha Venkatesh, and Dinh Phung. 2017. Latent sentiment topic modelling and nonparametric discovery of online mental health-related communities. *International Journal of Data Science and Analytics*, 4(3):209–231.

D. L. Davies and D. W. Bouldin. 1979. A cluster separation measure. *IEEE Transactions on Pattern Analysis and Machine Intelligence*, PAMI-1(2):224–227.

Munmun De Choudhury and Sushovan De. 2014. Mental health discourse on reddit: Self-disclosure, social support, and anonymity. *Proceedings of the International AAAI Conference on Web and Social Media*, 8(1).

Jacob Devlin, Ming-Wei Chang, Kenton Lee, and Kristina Toutanova. 2019. BERT: Pre-training of deep bidirectional transformers for language understanding. In *Proceedings of the 2019 Conference of the North American Chapter of the Association for Computational Linguistics: Human Language Technologies, Volume 1 (Long and Short Papers)*, pages 4171–4186, Minneapolis, Minnesota. Association for Computational Linguistics.

Elisa van Ee, Rolf J Kleber, Marian J Jongmans, Trudy TM Mooren, and Dorothee Out. 2016. Parental ptsd, adverse parenting and child attachment in a refugee sample. *Attachment & human development*, 18(3):273–291.

Zhengping Jiang, Sarah Ita Levitan, Jonathan Zomick, and Julia Hirschberg. 2020. Detection of mental health from Reddit via deep contextualized representations. In *Proceedings of the 11th International Workshop on Health Text Mining and Information Analysis*, pages 147–156, Online. Association for Computational Linguistics.

Diederik P. Kingma and Jimmy Ba. 2015. Adam: A method for stochastic optimization. In *3rd International Conference on Learning Representations, ICLR 2015, San Diego, CA, USA, May 7-9, 2015, Conference Track Proceedings*.

Prasadith Kirinde Gamaarachchige and Diana Inkpen. 2019. Multi-task, multi-channel, multi-input learning for mental illness detection using social media text. In *Proceedings of the Tenth International Workshop on Health Text Mining and Information Analysis (LOUHI 2019)*, pages 54–64, Hong Kong. Association for Computational Linguistics.

Jey Han Lau, David Newman, and Timothy Baldwin. 2014. Machine reading tea leaves: Automatically evaluating topic coherence and topic model quality. In *Proceedings of the 14th Conference of the European Chapter of the Association for Computational Linguistics*, pages 530–539, Gothenburg, Sweden. Association for Computational Linguistics.

Yinhan Liu, Myle Ott, Naman Goyal, Jingfei Du, Mandar Joshi, Danqi Chen, Omer Levy, Mike Lewis, Luke Zettlemoyer, and Veselin Stoyanov. 2019. Roberta: A robustly optimized bert pretraining approach. *arXiv preprint arXiv:1907.11692*.

Kate Loveys, Jonathan Torrez, Alex Fine, Glen Moriarty, and Glen Coppersmith. 2018. Cross-cultural differences in language markers of depression online. In *Proceedings of the Fifth Workshop on Computational Linguistics and Clinical Psychology: From Keyboard to Clinic*, pages 78–87, New Orleans, LA. Association for Computational Linguistics.

Andrew L. Maas, Awni Y. Hannun, and Andrew Y. Ng. 2013. Rectifier nonlinearities improve neural network acoustic models. In *in ICML Workshop on Deep Learning for Audio, Speech and Language Processing*.

Fiona Martin and Mark Johnson. 2015. More efficient topic modelling through a noun only approach. In *Proceedings of the Australasian Language Technology Association Workshop 2015*, pages 111–115, Parramatta, Australia.

Leland McInnes, John Healy, and Steve Astels. 2017. hdbscan: Hierarchical density based clustering. *The Journal of Open Source Software*, 2(11).

Leland McInnes, John Healy, and James Melville. 2018. Umap: Uniform manifold approximation and projection for dimension reduction. *arXiv preprint arXiv:1802.03426*.

David Mimno, Hanna Wallach, Edmund Talley, Miriam Leenders, and Andrew McCallum. 2011. Optimizing semantic coherence in topic models. In *Proceedings of the 2011 Conference on Empirical Methods in Natural Language Processing*, pages 262–272, Edinburgh, Scotland, UK. Association for Computational Linguistics.

Albert Park and Mike Conway. 2018. Harnessing reddit to understand the written-communication challenges experienced by individuals with mental health disorders: Analysis of texts from mental health communities. *J Med Internet Res*, 20(4):e121.

Albert Park, Mike Conway, and Annie T. Chen. 2018. Examining thematic similarity, difference, and membership in three online mental health communities from reddit. *Comput. Hum. Behav.*, 78(C):98–112.

Nicole Peinelt, Dong Nguyen, and Maria Liakata. 2020. tBERT: Topic models and BERT joining forces for semantic similarity detection. In *Proceedings of the 58th Annual Meeting of the Association for Computational Linguistics*, pages 7047–7055, Online. Association for Computational Linguistics.

Peter Rousseeuw. 1987. Silhouettes: A graphical aid to the interpretation and validation of cluster analysis. *J. Comput. Appl. Math.*, 20(1):53–65.

Hinrich Schütze, Christopher D Manning, and Prabhakar Raghavan. 2008. *Introduction to information retrieval*, volume 39. Cambridge University Press Cambridge.

Ivan Sekulic and Michael Strube. 2019. Adapting deep learning methods for mental health prediction on social media. In *Proceedings of the 5th Workshop on Noisy User-generated Text (W-NUT 2019)*, pages 322–327, Hong Kong, China. Association for Computational Linguistics.

Suzanna Sia, Ayush Dalmia, and Sabrina J. Mielke. 2020. Tired of topic models? clusters of pretrained word embeddings make for fast and good topics too! In *Proceedings of the 2020 Conference on Empirical Methods in Natural Language Processing (EMNLP)*, pages 1728–1736, Online. Association for Computational Linguistics.

B. Silveira Fraga, A. P. Couto da Silva, and F. Murai. 2018. Online social networks in health care: A study of mental disorders on reddit. In *2018 IEEE/WIC/ACM International Conference on Web Intelligence (WI)*, pages 568–573.

Keith Stevens, Philip Kegelmeyer, David Andrzejewski, and David Buttler. 2012. Exploring topic coherence over many models and many topics. In *Proceedings of the 2012 Joint Conference on Empirical Methods in Natural Language Processing and Computational Natural Language Learning*, pages 952–961, Jeju Island, Korea. Association for Computational Linguistics.

Laure Thompson and David Mimno. 2020. Topic modeling with contextualized word representation clusters. *arXiv preprint arXiv:2010.12626*.

Andrew Toulis and Lukasz Golab. 2017. Social media mining to understand public mental health. In *VLDB Workshop on Data Management and Analytics for Medicine and Healthcare*, pages 55–70. Springer.

Pascal Vincent, Hugo Larochelle, Isabelle Lajoie, Yoshua Bengio, and Pierre-Antoine Manzagol. 2010. Stacked denoising autoencoders: Learning useful representations in a deep network with a local denoising criterion. *Journal of Machine Learning Research*, 11(110):3371–3408.

Scientific Claim Verification with VERT5ERINI

Ronak Pradeep, Xueguang Ma, Rodrigo Nogueira, and Jimmy Lin
David R. Cheriton School of Computer Science
University of Waterloo

Abstract

This work describes the adaptation of a pretrained sequence-to-sequence model to the task of scientific claim verification in the biomedical domain. We propose a system called VERT5ERINI that exploits T5 for abstract retrieval, sentence selection, and label prediction, which are three critical sub-tasks of claim verification. We evaluate our pipeline on SCIFACT, a newly curated dataset that requires models to not just predict the veracity of claims but also provide relevant sentences from a corpus of scientific literature that support the prediction. Empirically, our system outperforms a strong baseline in each of the three sub-tasks. We further show VERT5ERINI's ability to generalize to two new datasets of COVID-19 claims using evidence from the CORD-19 corpus.

1 Introduction

The popularity of social media and other means of disseminating content, combined with automated algorithms that create "echo chamber" effects, has increased the proliferation of misinformation online. This has led to increased attention in the community on building better fact verification systems. Until recently, most fact verification datasets were constrained to domains such as Wikipedia, discussion blogs, and social media (Thorne et al., 2018; Hanselowski et al., 2019).

In the current environment, amidst the COVID-19 pandemic and the unease that comes with insufficient insight about the virus, there has been a sharp increase in curiosity among the general public toward scientific knowledge. While such curiosity is always appreciated, this has inadvertently resulted in a large spike of scientific facts being misrepresented, often to push a personal or political agenda, inducing ineffective and frequently even harmful policies and behaviours.

To mitigate this issue, Wadden et al. (2020) introduced the task of scientific claim verification, where systems need to evaluate the veracity of a claim against a scientific corpus. To facilitate this, they introduced the SCIFACT dataset that consists of scientific claims accompanied with abstracts that either support or refute the claim. The dataset also provides a set of rationale sentences for each claim that is necessary and sufficient to conclude its veracity. In addition, the authors provide VERISCI, a baseline for this task that takes inspiration from previous state-of-the-art systems (DeYoung et al., 2020) for the FEVER claim verification dataset (Thorne et al., 2018). This pipeline retrieves relevant abstracts by TF-IDF similarity, uses a BERT-based model (Devlin et al., 2019) to select rationale sentences, and finally labels each abstract as either SUPPORTS, NOINFO, or REFUTES with respect to the claim.

Despite the success of BERT for tasks like passage-level (Nogueira et al., 2019), document-level (Dai and Callan, 2019; MacAvaney et al., 2019; Akkalyoncu Yilmaz et al., 2019) and sentence-level (Soleimani et al., 2019) retrieval, there is evidence that ranking with sequence-to-sequence models can achieve even better effectiveness, particularly in zero-shot scenarios or with limited training data (Nogueira et al., 2020; Pradeep et al., 2021). This was further demonstrated in the TREC-COVID challenge (Roberts et al., 2020) where one of the best performing systems used sequence-to-sequence models for retrieval (Zhang et al., 2020; Pradeep et al., 2021). Similar trends are noted in CovidQA (Tang et al., 2020), a question answering dataset for COVID-19, where zero-shot sequence-to-sequence models outperformed other baselines.

Hence, we propose VERT5ERINI, where all three steps—abstract retrieval, sentence selection, and label prediction exploit T5 (Raffel et al., 2020),

Proceedings of the 12th International Workshop on Health Text Mining and Information Analysis, pages 94–103
April 19, 2021. ©2021 Association for Computational Linguistics

a powerful sequence-to-sequence language model. VERT5ERINI outperforms the VERISCI baseline on the SCIFACT tasks by a large margin and advances the state of the art for the task of Scientific Claim Verification. We also demonstrate the effectiveness of our system in verifying two different sets of COVID-19 claims with no additional training or hyperparameter tuning.

2 Task

In the SCIFACT task (Wadden et al., 2020), systems are provided with a scientific claim q and a corpus of abstracts $\mathcal{C}$ and tasked to return:

- A set of evidence abstracts $\hat{\mathcal{E}}(q)$.
- A label $\hat{y}(q, a)$ that maps claim q and abstract a to one of $\{\text{SUPPORTS}, \text{REFUTES}, \text{NOINFO}\}$.
- A set of rationale sentences $\hat{S}(q, a)$ when $\hat{y}(q, a) \in \{\text{SUPPORTS}, \text{REFUTES}\}$.

Given the ground truth label $y(q, a)$, the set of gold abstracts $\mathcal{E}(q)$, and the set of gold rationales $\mathcal{R}(q, a)$ (each gold rationale is a set of sentences), the predictions are evaluated in two ways:

- **Abstract-level evaluation**, where systems are judged on whether they can identify abstracts that support or refute the claim. First, $a \in \hat{\mathcal{E}}(q)$ is *correctly labelled* if both $a \in \mathcal{E}(q)$ and $\hat{y}(q, a) = y(q, a)$. Second, it is *correctly rationalized*, if in addition, $\exists R \in \mathcal{R}(q, a)$ such that $R \subseteq \hat{S}(q, a)$.[1] These evaluations are referred to as $\text{Abstract}_{\text{Label-Only}}$ and $\text{Abstract}_{\text{Label+Rationale}}$, respectively.
- **Sentence-level evaluation**, where systems are evaluated on whether they can identify sentences sufficient to justify the abstract-level predictions. First, $\hat{s} \in \hat{S}(q, a)$ is *correctly selected* if $\exists R \in \mathcal{R}(q, a)$ such that both $\hat{s} \in R$ and $R \subseteq \hat{S}(q, a)$. Second, it is *correctly labelled*, if in addition, $\hat{y}(q, a) = y(q, a)$. These evaluations are referred to as $\text{Sentence}_{\text{Selection-Only}}$ and $\text{Sentence}_{\text{Selection+Label}}$, respectively.

Specifically, SCIFACT uses a corpus of 5,183 abstracts. Abstracts that support or refute each claim are annotated with rationale sentences (see Table 3 for examples). The label distribution is provided in

[1]In SCIFACT's abstract-level evaluation, it is required that $|\hat{S}(q, a)| \leq 3$.

Set	SUPPORTS	NOINFO	REFUTES	Total
Train	332	304	173	809
Dev	124	112	64	300
Test	100	100	100	300

Table 1: SCIFACT label distribution.

Claim Set	SUPPORTS	REFUTES	Total
COVID-19 SCIFACT	-	-	36
COVID-19 Scientific	41	101	142

Table 2: COVID-19 claims.

Table 1. There are 1,409 claims, 809 of which are part of the training set and the rest are split equally across the development and test sets. Although the test set is balanced with 100 claims for each class (SUPPORTS, NOINFO, and REFUTES), it is clear that the training and development sets have significant class imbalance. This, coupled with the small dataset size, highlights the importance of zero- or few-shot systems for this task.

To show that our system is able to verify claims related to COVID-19 by identifying evidence from the much larger CORD-19 corpus,[2] we evaluate VERT5ERINI in a zero-shot setting on two other datasets:

COVID-19 SCIFACT (Wadden et al., 2020) is a set of 36 COVID-related claims curated by a medical student. In this set, the same claim can sometimes be both supported and refuted by different abstracts, a scenario not observed in the main SCIFACT task. Two examples in this set are shown in Table 4.

COVID-19 Scientific (Lee et al., 2020) contains 142 claims (label distribution in Table 2) gathered by collecting COVID-related scientific truths and myths from sources like the U.S. Centers for Disease Control and Prevention (CDC), MedicalNewsToday, and the World Health Organization (WHO). Unlike the other two datasets, COVID-19 Scientific only provides a single label $y(q) \in \{\text{SUPPORTS}, \text{REFUTES}\}$ for a claim. According to the authors, during the construction of the dataset, claims that were unverifiable according to the CDC or the WHO were mapped to REFUTES. Hence, we make the following modifications to VERT5ERINI:

1. If $\hat{y}(q, a) = \text{NOINFO}$, then $\hat{y}(q, a)$ is modified to REFUTES.

[2]We use the 2020-06-17 dump of CORD-19, which contains 192,459 abstracts, about 40 times as many abstracts as those in SCIFACT.

Claim	Label	Evidence
ALDH1 expression is associated with poorer prognosis in breast cancer.	SUPPORTS	Application of stem cell biology to breast cancer research has been limited by the lack of simple methods for identification and isolation of normal and malignant stem cells. ... **In a series of 577 breast carcinomas, expression of ALDH1 detected by immunostaining correlated with poor prognosis.** ...
CX3CR1 on the Th2 cells impairs T cell survival	REFUTES	Allergic asthma is a T helper type 2 (T(H)2)-dominated disease of the lung. ... **We found that CX3CR1 signaling promoted T(H)2 survival in the inflamed lungs, and injection of B cell leukemia/lymphoma-2 protein (BCl-2)-transduced CX3CR1-deficient T(H)2 cells into CX3CR1-deficient mice restored asthma.** ...
Arterioles have a larger lumen diameter than venules.	NOINFO	N/A

Table 3: Three SCIFACT claims, their labels, and their corresponding evidence (rationale highlighted in **bold**) if available. VERT5ERINI correctly predicts these labels and retrieves the matching evidence.

Claim	Label	Evidence
Hypertension and Diabetes are the most common comorbidities for COVID-19.	SUPPORTS	**Investigations reported that hypertension, diabetes, and cardiovascular diseases were the most prevalent comorbidities among the patients with coronavirus disease 2019 (COVID-19).** ... The aim of this review was to summarize the current knowledge about the relationship between hypertension and COVID-19 and the role of hypertension on outcome in these patients.
The Secondary Attack rate of COVID-19 is 10.5% for household members/close contacts.	REFUTES	Background: As of April 2, 2020, the global reported number of COVID-19 cases has crossed over 1 million with more than 55,000 deaths. ... **We estimated the household SAR to be 13.8% (95% CI: 11.1–17.0%) if household contacts are defined as all close relatives and 19.3% (95% CI: 15.5–23.9%) if household contacts only include those at the same residential address as the cases, assuming a mean incubation period of 4 days and a maximum infectious period of 13 days.**

Table 4: Two COVID-19 claims from SCIFACT, their *predicted* labels and their corresponding *predicted* evidence (rationale highlighted in **bold**).

Claim	Label	Evidence
Some people become infected by COVID-19 but don't develop any symptoms and don't feel unwell.	SUPPORTS	COVID-19 is an emerging infectious disease with widespread transmission of the coronavirus SARS-CoV-2 in the Netherlands. ... **Others do not show any symptoms, but can still contribute to the transmission of the virus.** ...
Young people will not get COVID-19.	REFUTES	Objective: To explore the epidemiological characteristics of COVID-19 associated with SARS-Cov-2 in Guizhou province, and to compare the differences in epidemiology with other provinces. ... **Most of COVID-19 patients were 18-45 years old (52.27%).** ... **CONCLUSION: Among the cases, most patients were young adults.**
Bill Gates caused the infection of COVID-19.	REFUTES	N/A

Table 5: Three COVID-19 Scientific claims from Lee et al. (2020), their *predicted* labels and their corresponding *predicted* evidence (rationale highlighted in **bold**). Note that if the system cannot find any supporting evidence for a claim, it is considered refuted.

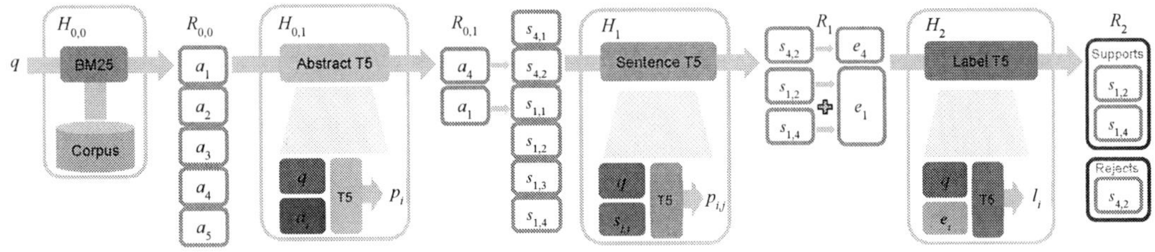

Figure 1: Illustration of the VERT5ERINI pipeline. In stage $H_{0,0}$, given a query q, the top $k_{0,0}$ ($= 5$ in the figure) candidate documents $R_{0,0}$ are retrieved using BM25. In stage $H_{0,1}$, Abstract T5 produces a relevance score p_i for each pair of query q and candidate $a_i \in R_{0,0}$. The top $k_{0,1}$ ($= 2$ in the figure) candidates with respect to these relevance scores are expanded to sentence-level granularity and passed to stage H_1, in which Sentence T5 computes a relevance score $p_{i,j}$ for each pair of query q and candidate sentence $s_{i,j}$. Sentences for a particular abstract a_i scoring above a threshold form its set of rationale sentences e_i and each set along with the query q are passed to stage H_2, in which Label T5 predicts the label.

2. $\hat{y}(q) = \max_{a \in \hat{\mathcal{E}}(q)} \hat{y}(q, a)$.

3. If $|\bigcup_{a \in \hat{\mathcal{E}}(q)} \hat{S}(q, a)| = 0$, i.e., the set of all evidence sentences across the abstracts is empty, then $\hat{y}(q) =$ REFUTES.

Three examples from this set are shown in Table 5. As one can imagine, it would be impossible to find any discussion of outlandish claims like "Bill Gates caused the infection of COVID-19" in a corpus of biomedical literature and hence VERT5ERINI maps them to REFUTES.

3 Methods

Our proposed system, VERT5ERINI (see Figure 1), has three major components:

1. H_0: **Abstract Retrieval** — which given claim q retrieves the top-k abstracts from corpus $\mathcal{C}$.

2. H_1: **Sentence Selection** — which given claim q and one of the top-k abstracts a, selects sentences from a that form $\hat{S}(q, a)$.

3. H_2: **Label Prediction** — which given claim q and the rationale sentences $\hat{S}(q, a)$, predicts the final label $\hat{y}(q, a)$.

3.1 H_0: Abstract Retrieval

Given a scientific claim q and a corpus $\mathcal{C}$ of scientific abstracts, H_0 is tasked with retrieving the top-k abstracts from $\mathcal{C}$. We propose both a single-stage and a two-stage abstract retrieval pipeline.

In both cases, the first stage $H_{0,0}$ involves treating the query as a "bag of words" for ranking abstracts from the corpus using the BM25 scoring function (Robertson et al., 1994). Our implementation uses the Anserini IR toolkit (Yang et al., 2017, 2018),[3] which is built on the popular open-source Lucene search engine, and its Pyserini Python interface (Akkalyoncu Yilmaz et al., 2020; Lin et al., 2021) to support simple keyword search capabilities on the corpus. The output of this stage is a list of k_0 candidate abstracts.

The second abstract reranking stage, $H_{0,1}$, is tasked to estimate a score p quantifying how relevant a candidate abstract a is to a query q. In this stage, the abstracts retrieved in $H_{0,0}$ are reranked by a pointwise reranker, which we call monoT5. Our reranker is based on Nogueira et al. (2020), which uses T5 (Raffel et al., 2020), a sequence-to-sequence model pretrained with a similar masked language modeling objective as BERT. In this model, all target tasks are cast as sequence-to-sequence tasks. We adapt the approach to abstract reranking by using the following input sequence:

Query: q Document: a Relevant:

The model is fine-tuned to produce the words "true" or "false" depending on whether the abstract is relevant or not to the query. That is, "true" and "false" are the "target words" (i.e., ground truth predictions in the sequence-to-sequence transformation).

[3] `http://anserini.io/`

Since a considerable number ($\approx 15\%$) of SCIFACT abstracts are longer than the context limit of T5 (512 tokens), we first segment each abstract (on average 9 sentences) into spans by applying a sliding window of 6 sentences with a stride of 3.

In order to fine-tune monoT5 on abstract reranking in SCIFACT, we use all cited abstracts in the training set as positive examples. For each claim, we select negative examples by randomly selecting a non-ground truth abstract among the top-10 BM25 ranked candidates. We train on this set with a batch size of 128 for 200 steps, which corresponds to approximately 5 epochs.

At inference time, we first compute probabilities for each query–segment pair (in a reranking setting) by applying a softmax only on the logits of the "true" and "false" tokens. We then obtain the relevance score of the document as the highest probability assigned to the "true" token among all segments. The top-k_0 abstracts, R_0, with respect to these scores are then selected.

We run inference with three different monoT5[4] settings for abstract reranking: (1) fine-tuned on the MS MARCO passage dataset (Bajaj et al., 2016); (2) fine-tuned on MS MARCO and then fine-tuned again on the medical subset of MS MARCO (MacAvaney et al., 2020); and (3) fine-tuned on MS MARCO and then fine-tuned again on SCIFACT.

We choose to "pre–fine-tune" relevance classifiers on MS MARCO passages as it has been shown to help in various other tasks (Akkalyoncu Yilmaz et al., 2019; Zhang et al., 2020; Nogueira et al., 2020; Pradeep et al., 2021). Similarly, MacAvaney et al. (2020) demonstrated that fine-tuning the classifiers on the medical subset of MS MARCO helps with biomedical-domain relevance ranking.

3.2 H_1: Sentence Selection

In this stage, the goal is to select rationale sentences $\hat{S}(q, a)$ from each abstract a for each of the top-k abstracts retrieved $\hat{\mathcal{E}}(q)$. We use T5 for this task also. The following input sequence is used:

Query: q Document: s Relevant:

where s is a sentence in the abstract a.

We fine-tune a monoT5 (pre–fine-tuned on MS MARCO passage) on SCIFACT's gold rationales as positive examples and sentences randomly sampled from $\mathcal{E}(q)$ as negatives. We train on this set of sentences with a batch size of 128 for 2500 steps.

During inference, similar to abstract ranking, we compute a probability of the sentence being relevant based on the logits of the "true" and "false" tokens. Finally, we filter out all sentences whose "true" probability is below the threshold of 0.999 to obtain $\hat{S}(q, a)$.

3.3 H_2: Label Prediction

Given the claim q, an abstract a and their corresponding set of rationale sentences $\hat{S}(q, a)$, H_2 is tasked to predict a label $\hat{y}(q, a) \in \{\text{SUPPORTS}, \text{NOINFO}, \text{REFUTES}\}$. Yet again, we use T5 for this task with the input sequence:

hypothesis: q sentence1: s_1 $\cdots$ sentence$z$: s_z

where $s_1, \cdots, s_z$ are the rationale sentences in $\hat{S}(q, a)$. The target sequence is one of "true", "weak", or "false" tokens corresponding to the labels SUPPORTS, NOINFO, or REFUTES, respectively. Note that a feature of this approach is feeding a collection of sentences into the model at once, as opposed to the perhaps more obvious approach of performing per-sentence independent label prediction. This requires the model to process longer input sequences, but allows predictions to incorporate evidence from multiple sources. In the parlance of learning to rank in the context of information retrieval, this would be called a "listwise" approach (Li, 2011).

SUPPORTS and REFUTES training examples are selected from evidence sets of cited abstracts for each claim. The sentences in each evidence set are concatenated with the claim in the above input sequence template as a single example for the corresponding label. The NOINFO examples are selected by concatenating one or two randomly-selected non-rationale sentences from each of the cited abstracts across all labels. Here, we fine-tune a fresh T5-3B (that was just pretrained on the mixture task) and not a pre–fine-tuned monoT5 since there is no natural transfer from the relevance ranking task. We use a batch size of 128 and select the best checkpoint after $[200, 400, 600, 800, 1000]$ steps based on the development set scores.

During inference, the token with the highest probability is assigned the label $\hat{y}(q, a)$ for abstract $a \in \hat{\mathcal{E}}(q)$.

[4]All models are T5-3B.

Method	R@3	R@5
Oracle	97.61	100.00
TF-IDF	69.38	75.60
BM25	79.90	84.69
T5 (MS MARCO)	86.12	**89.95**
T5 (MS MARCO MED)	85.65	89.00
T5 (SCIFACT)	**86.60**	89.40

Table 6: Comparison of abstract retrieval methods on the development set of SCIFACT.

4 Results

4.1 Baselines

For the SCIFACT and COVID-19 SCIFACT end-to-end tasks, the baseline system used is VERISCI (Wadden et al., 2020). It has an abstract retrieval module that uses TF-IDF, a sentence selection module trained on SCIFACT, and a label prediction module trained on FEVER + SCIFACT. For the abstract retrieval module, the authors report the best full-pipeline development set scores by retrieving the top three documents.

For the COVID-19 Scientific task, we compare with the following two baselines established by Lee et al. (2020):

- LiarMisinfo (Lee et al., 2020) uses a BERT-large (Devlin et al., 2019) label prediction model fine-tuned on LIAR-PolitiFact (Wang, 2017), a set of 12.8k claims collected from PolitiFact. It is worth noting that LIAR-PolitiFact does not contain any claims related to COVID-19.

- LM Debunker (Lee et al., 2020) uses GPT-2 (Radford et al., 2019) to determine the perplexity of the claim given evidence sentences. Claims with a perplexity score higher than a threshold are labeled REFUTES while the others are labeled SUPPORTS.

The sentence selection module in both baselines employ TF-IDF followed by some rule-based evidence filtering to select the top three sentences for each claim. LiarMisinfo represents a zero-shot model where no fine-tuning is performed on the COVID-19 Scientific set. LM Debunker, on the other hand, first partitions the data into a validation and a test set. The validation set is used to tune the perplexity threshold for the model following which evaluation is performed on the test set.

Method	P	R	F_1
RoBERTa-large	73.71	70.49	72.07
T5	**79.29**	**73.22**	**76.14**

Table 7: Comparison of different sentence selection methods on SCIFACT's development set.

Method	Label	P	R	F_1
RoBERTa-large	SUPPORTS	92.56	81.16	86.49
	NOINFO	74.82	**92.86**	82.87
	REFUTES	77.05	66.20	71.21
T5	SUPPORTS	**93.13**	**88.41**	**90.71**
	NOINFO	**85.25**	**92.86**	**88.89**
	REFUTES	**86.76**	**83.10**	**84.89**

Table 8: Comparison of different label prediction models on SCIFACT's development set.

4.2 Abstract Retrieval

Table 6 reports recall at rank three (R@3) and rank five (R@5) for abstract retrieval. The oracle (first row) shows that most claims from the development set have fewer than three relevant abstracts and all have fewer than five. For comparison, we show the effectiveness of the TF-IDF method used by Wadden et al. (2020).

We find that using BM25 results in an effectiveness improvement of around 10 points in comparison to the TF-IDF baseline. Using T5 to rerank the top-20 abstracts retrieved from BM25 results in a 17-point improvement over the baseline.

However, results show almost no difference in effectiveness whether T5 was fine-tuned on SCIFACT or on MS MARCO MED. This might be due to the relatively small size of the SCIFACT dataset and the fact that MS MARCO MED data is not entirely relevant to the target task. Hence, we use T5 fine-tuned only on the full MS MARCO dataset (i.e., no further fine-tuning) in the end-to-end pipeline experiments (Section 4.5).

4.3 Sentence Selection

Table 7 reports the precision, recall, and F_1 scores for the sentence selection task. We find that T5 (MS MARCO) fine-tuned on SCIFACT outperforms the RoBERTa-large baseline fine-tuned on SCIFACT used by Wadden et al. (2020). This result, together with results from Table 6, demonstrates the effectiveness of the T5 model at selecting evidence at various levels of granularity.

Label Only			
Method	P	R	F_1
(1) Oracle (VERISCI)	90.97	67.46	77.47
(2) Oracle (ours)	**92.70**	**78.95**	**85.27**
(3) VERISCI	55.31	47.37	51.03
(4) VERT5ERINI (BM25)	**70.88**	61.72	**65.98**
(5) VERT5ERINI (T5)	65.07	**65.07**	65.07
Label+Rationale			
Method	P	R	F_1
(6) Oracle (VERISCI)	85.16	63.16	72.53
(7) Oracle (ours)	**88.76**	**75.60**	**81.65**
(8) VERISCI	52.51	44.98	48.45
(9) VERT5ERINI (BM25)	**67.03**	58.37	**62.40**
(10) VERT5ERINI (T5)	61.72	**61.72**	61.72

Table 9: Full pipeline abstract-level effectiveness on SCIFACT's development set.

Selection Only			
Method	P	R	F_1
(1) Oracle (VERISCI)	79.41	59.02	67.71
(2) Oracle (ours)	**83.54**	**72.13**	**77.42**
(3) VERISCI	52.46	43.72	47.69
(4) VERT5ERINI (BM25)	**67.70**	53.83	59.97
(5) VERT5ERINI (T5)	64.81	**57.37**	**60.87**
Selection+Label			
Method	P	R	F_1
(6) Oracle (VERISCI)	71.32	53.01	60.82
(7) Oracle (ours)	**78.16**	**67.49**	**72.43**
(8) VERISCI	46.89	39.07	42.62
(9) VERT5ERINI (BM25)	**63.92**	50.82	56.62
(10) VERT5ERINI (T5)	60.80	**53.83**	**57.10**

Table 10: Full pipeline sentence-level effectiveness on SCIFACT's development set.

4.4 Label Prediction

In Table 8, we present label-wise precision, recall, and F_1 scores for the label prediction task. For SUPPORTS and REFUTES labels, the input to the model comprises gold rationales from cited abstracts. For NOINFO labels, recall that cited abstracts are available but no gold rationales exist. In this case, we pick the two most similar sentences according to TF-IDF from each abstract.

The results across all labels demonstrate that T5 fine-tuned on SCIFACT's label prediction task shows significant improvements over the baseline RoBERTa-large that was fine-tuned on FEVER followed by fine-tuning on SCIFACT's label prediction task. We believe some of this can be credited to T5's pretraining on a mixture of multiple tasks. Although this mixture does not include FEVER, the corpus contains various other NLI datasets, including MNLI (Williams et al., 2018) and QNLI (Rajpurkar et al., 2016).

4.5 Full Pipeline

In Tables 9 and 10, we report the precision, recall, and F_1 scores of abstract-level evaluation and sentence-level evaluation, respectively, for full pipeline systems.

Rows 1, 2, 6, 7 present the scores in the oracle abstract retrieval setting, where gold evidence abstracts are provided to systems. We see that our pipeline outperforms VERISCI by around 10 F_1 points at both the abstract and sentence level. The improvements are even larger in the $\text{Abstract}_{\text{Label+Rationale}}$ and $\text{Sentence}_{\text{Selection+Label}}$ evaluation settings (rows 6, 7 in Tables 9 and 10, respectively) which require more from systems in terms of sentence selection and label prediction.

In rows 3–5 and 8–10, we report scores in the full pipeline setting where systems are also required to retrieve relevant abstracts. We evaluate two full pipeline systems, one that uses BM25 alone and another that uses BM25 followed by T5 (MARCO) for abstract retrieval. Both these systems outperform the baseline system VERISCI by about 14 F_1 points. This comes as no surprise seeing that our models display notable improvements for each of the three sub-tasks.

Notice that in Table 6, using T5 (MARCO) brings large gains in terms of R@3 over the BM25 baseline. Yet, in the case of the full pipeline, with these two abstract retrieval methods, we only observe comparable effectiveness on the development set. We believe this might be linked to the relatively small size of the development set; below, we choose to probe the SCIFACT hidden test set with both configurations.

From Table 11, it is clear that in the hidden test set, both our systems outperform the baseline VERISCI, with evaluation aspects like Sentence+Label (rows 10–12) showing relative improvements of around 50%. Comparing with the corresponding conditions in Tables 9 and 10, we also see no indication of overfitting. We also note that abstract retrieval using the two-stage approach brings large gains here (rows 5, 11 vs. 6, 12) unlike in the development set. This shows that neural reranking, even though used in a zero-shot formu-

Label Only			
Method	P	R	F_1
(1) VERISCI	47.5	47.3	47.4
(2) VERT5ERINI (BM25)	63.1	60.8	61.9
(3) VERT5ERINI (T5)	**63.6**	**66.2**	**64.9**
Label+Rationale			
Method	P	R	F_1
(4) VERISCI	46.6	46.4	46.5
(5) VERT5ERINI (BM25)	60.3	58.1	59.2
(6) VERT5ERINI (T5)	**61.5**	**64.0**	**62.7**
Selection Only			
Method	P	R	F_1
(7) VERISCI	45.0	47.3	46.1
(8) VERT5ERINI (BM25)	64.9	58.9	61.8
(9) VERT5ERINI (T5)	**66.2**	**63.5**	**64.8**
Selection+Label			
Method	P	R	F_1
(10) VERISCI	38.6	40.5	39.5
(11) VERT5ERINI (BM25)	58.3	53.0	55.5
(12) VERT5ERINI (T5)	**60.0**	**57.6**	**58.8**

Table 11: Full pipeline effectiveness of VERT5ERINI on SCIFACT's test set.

lation, is critical to getting higher quality abstracts from the corpus C, thereby improving effectiveness in later stages too.

4.6 Verification of COVID-19 Claims

Finally, we evaluate our most effective pipeline configuration, VERT5ERINI (T5), on the two sets of COVID-related claims. We do this in a zero-shot setting in that we do not fine-tune our model on either of these datasets.

In the COVID-19 SCIFACT set, for each claim q, we use VERT5ERINI (T5) to predict evidence abstracts, $\hat{\mathcal{E}}(q)$. A $(q, \hat{\mathcal{E}}(q))$ pair is considered *plausible* if at least half of the evidence abstracts in $\hat{\mathcal{E}}(q)$ are found to have reasonable rationales and labels. For 30 out of 36 claims, we find that VERT5ERINI (T5) provides plausible evidence abstracts. These claims have reasonable labels and evidence rationales selected successfully from evidence abstracts. This is in comparison to the 23 out of 36 claims for which VERISCI provides plausible evidence, demonstrating the effectiveness of our system in the zero-shot setting.

In the COVID-19 Scientific set, we compare the effectiveness of VERT5ERINI with that of two baselines considered by Lee et al. (2020). Table 12 reports the accuracy, the F_1-Macro, and the F_1-Binary scores on the test set. The F_1-Binary score corresponds to the F_1 score of the REFUTES label, since debunking misinformation is critical. Note that the LM Debunker baseline uses the average scores across four-fold cross-validation on the test set, unlike VERT5ERINI and LiarMisinfo. We observe that VERT5ERINI outperforms both baselines in a zero-shot setting, without any in-task tuning like the LM Debunker. The adaptability of VERT5ERINI to both these new tasks with no additional training makes a strong case for the effectiveness of our system.

Method	Accuracy	F_1-Macro	F_1-Binary
LiarMisinfo	61.5	59.2	82.8
LM Debunker	75.4	69.8	83.1
VERT5ERINI (T5)	**78.2**	**73.2**	**83.8**

Table 12: Label prediction effectiveness on COVID-19 Scientific claims

5 Conclusions

In this paper, we introduced VERT5ERINI, a novel system for scientific claim verification that exploits a generation-based approach to abstract ranking, sentence selection, and claim verification. Such systems are of significance in this age of misinformation, amplified by the COVID-19 pandemic. Experiments show that our system outperforms the state of the art in the end-to-end task on the SCIFACT dataset. We note improvements in each of the three sub-tasks, demonstrating the importance of this sequence-to-sequence approach as well as zero-shot and few-shot transfer capabilities. Finally, we find that VERT5ERINI generalizes to two new COVID-19 related datasets with no tuning of parameters while maintaining high effectiveness.

Yet, there is still a large gap between our system and an oracle. Ideally, a system that performs scientific claim verification should possess additional attributes such as:

- Numerical reasoning — the ability to interpret statistical and numerical findings and ranges.
- Biomedical background — the ability to leverage knowledge about domain-specific lexical relationships.

Future work that incorporates such attributes might be critical towards building higher-quality scientific fact verification systems. We report progress, but there is much more work to be done.

Acknowledgements

This research was supported in part by the Canada First Research Excellence Fund, the Natural Sciences and Engineering Research Council (NSERC) of Canada, and the Waterloo–Huawei Joint Innovation Laboratory. Additionally, we would like to thank Google for computational resources in the form of Google Cloud credits.

References

Zeynep Akkalyoncu Yilmaz, Charles L. A. Clarke, and Jimmy Lin. 2020. A lightweight environment for learning experimental IR research practices. In *Proceedings of the 43rd Annual International ACM SIGIR Conference on Research and Development in Information Retrieval (SIGIR 2020)*, pages 2113–2116.

Zeynep Akkalyoncu Yilmaz, Wei Yang, Haotian Zhang, and Jimmy Lin. 2019. Cross-domain modeling of sentence-level evidence for document retrieval. In *Proceedings of the 2019 Conference on Empirical Methods in Natural Language Processing and the 9th International Joint Conference on Natural Language Processing (EMNLP-IJCNLP)*, pages 3481–3487.

Payal Bajaj, Daniel Campos, Nick Craswell, Li Deng, Jianfeng Gao, Xiaodong Liu, Rangan Majumder, Andrew McNamara, Bhaskar Mitra, Tri Nguyen, Mir Rosenberg, Xia Song, Alina Stoica, Saurabh Tiwary, and Tong Wang. 2016. MS MARCO: A human generated MAchine Reading COmprehension dataset. *arXiv:1611.09268*.

Zhuyun Dai and Jamie Callan. 2019. Context-aware sentence/passage term importance estimation for first stage retrieval. *arXiv:1910.10687*.

Jacob Devlin, Ming-Wei Chang, Kenton Lee, and Kristina Toutanova. 2019. BERT: Pre-training of deep bidirectional transformers for language understanding. In *Proceedings of the 2019 Conference of the North American Chapter of the Association for Computational Linguistics: Human Language Technologies, Volume 1 (Long and Short Papers)*, pages 4171–4186, Minneapolis, Minnesota.

Jay DeYoung, Sarthak Jain, Nazneen Fatema Rajani, Eric Lehman, Caiming Xiong, Richard Socher, and Byron C. Wallace. 2020. ERASER: A benchmark to evaluate rationalized NLP models. In *Proceedings of the 58th Annual Meeting of the Association for Computational Linguistics*, pages 4443–4458.

Andreas Hanselowski, Christian Stab, Claudia Schulz, Zile Li, and Iryna Gurevych. 2019. A richly annotated corpus for different tasks in automated fact-checking. In *Proceedings of the 23rd Conference on Computational Natural Language Learning (CoNLL)*, pages 493–503, Hong Kong, China.

Nayeon Lee, Yejin Bang, Andrea Madotto, and Pascale Fung. 2020. Misinformation has high perplexity. *arXiv:2006.04666*.

Hang Li. 2011. *Learning to Rank for Information Retrieval and Natural Language Processing*. Morgan & Claypool Publishers.

Jimmy Lin, Xueguang Ma, Sheng-Chieh Lin, Jheng-Hong Yang, Ronak Pradeep, and Rodrigo Nogueira. 2021. Pyserini: An easy-to-use Python toolkit to support replicable IR research with sparse and dense representations. *arXiv:2102.10073*.

Sean MacAvaney, Arman Cohan, and Nazli Goharian. 2020. SLEDGE: A simple yet effective baseline for COVID-19 scientific knowledge search. *arXiv:2005.02365*.

Sean MacAvaney, Andrew Yates, Arman Cohan, and Nazli Goharian. 2019. CEDR: Contextualized embeddings for document ranking. In *Proceedings of the 42nd International ACM SIGIR Conference on Research and Development in Information Retrieval*, pages 1101–1104.

Rodrigo Nogueira, Zhiying Jiang, Ronak Pradeep, and Jimmy Lin. 2020. Document ranking with a pretrained sequence-to-sequence model. In *Findings of EMNLP*.

Rodrigo Nogueira, Wei Yang, Jimmy Lin, and Kyunghyun Cho. 2019. Document expansion by query prediction. *arXiv:1904.08375*.

Ronak Pradeep, Rodrigo Nogueira, and Jimmy Lin. 2021. The expando-mono-duo design pattern for text ranking with pretrained sequence-to-sequence models. *arXiv:2101.05667*.

Alec Radford, Jeffrey Wu, Rewon Child, David Luan, Dario Amodei, and Ilya Sutskever. 2019. Language models are unsupervised multitask learners. *OpenAI Technical Report*.

Colin Raffel, Noam Shazeer, Adam Roberts, Katherine Lee, Sharan Narang, Michael Matena, Yanqi Zhou, Wei Li, and Peter J. Liu. 2020. Exploring the limits of transfer learning with a unified text-to-text transformer. *Journal of Machine Learning Research*, 21(140):1–67.

Pranav Rajpurkar, Jian Zhang, Konstantin Lopyrev, and Percy Liang. 2016. SQuAD: 100,000+ questions for machine comprehension of text. In *Proceedings of the 2016 Conference on Empirical Methods in Natural Language Processing*, pages 2383–2392, Austin, Texas.

Kirk Roberts, Tasmeer Alam, Steven Bedrick, Dina Demner-Fushman, Kyle Lo, Ian Soboroff, Ellen Voorhees, Lucy Lu Wang, and William R. Hersh. 2020. TREC-COVID: Rationale and structure of an information retrieval shared task for COVID-19. *Journal of the American Medical Informatics Association*.

Stephen E. Robertson, Steve Walker, Susan Jones, Micheline Hancock-Beaulieu, and Mike Gatford. 1994. Okapi at TREC-3. In *Proceedings of the 3rd Text REtrieval Conference (TREC-3)*, pages 109–126.

Amir Soleimani, Christof Monz, and Marcel Worring. 2019. BERT for Evidence Retrieval and Claim Verification. *arXiv:1910.02655*.

Raphael Tang, Rodrigo Nogueira, Edwin Zhang, Nikhil Gupta, Phuong Cam, Kyunghyun Cho, and Jimmy Lin. 2020. Rapidly bootstrapping a question answering dataset for COVID-19. *arXiv:2004.11339*.

James Thorne, Andreas Vlachos, Christos Christodoulopoulos, and Arpit Mittal. 2018. FEVER: A large-scale dataset for Fact Extraction and VERification. In *Proceedings of the 2018 Conference of the North American Chapter of the Association for Computational Linguistics: Human Language Technologies, Volume 1 (Long Papers)*, pages 809–819, New Orleans, Louisiana.

David Wadden, Shanchuan Lin, Kyle Lo, Lucy Lu Wang, Madeleine van Zuylen, Arman Cohan, and Hannaneh Hajishirzi. 2020. Fact or fiction: Verifying scientific claims. In *Proceedings of the 2020 Conference on Empirical Methods in Natural Language Processing (EMNLP)*, pages 7534–7550.

William Yang Wang. 2017. "Liar, Liar Pants on Fire": A new benchmark dataset for Fake News Detection. *arXiv:1705.00648*.

Adina Williams, Nikita Nangia, and Samuel Bowman. 2018. A broad-coverage challenge corpus for sentence understanding through inference. In *Proceedings of the 2018 Conference of the North American Chapter of the Association for Computational Linguistics: Human Language Technologies, Volume 1 (Long Papers)*, pages 1112–1122, New Orleans, Louisiana.

Peilin Yang, Hui Fang, and Jimmy Lin. 2017. Anserini: Enabling the use of Lucene for information retrieval research. In *Proceedings of the 40th Annual International ACM SIGIR Conference on Research and Development in Information Retrieval (SIGIR 2017)*, pages 1253–1256, Tokyo, Japan.

Peilin Yang, Hui Fang, and Jimmy Lin. 2018. Anserini: Reproducible ranking baselines using Lucene. *Journal of Data and Information Quality*, 10(4):Article 16.

Edwin Zhang, Nikhil Gupta, Raphael Tang, Xiao Han, Ronak Pradeep, Kuang Lu, Yue Zhang, Rodrigo Nogueira, Kyunghyun Cho, Hui Fang, and Jimmy Lin. 2020. Covidex: Neural ranking models and keyword search infrastructure for the COVID-19 Open Research Dataset. *arXiv:2007.07846*.

Association for Computational Linguistics
209 N. Eighth Street
Stroudsburg, Pennsylvania 18360

ISBN 978-1-7138-2969-0

CURRAN ASSOCIATES INC.
proceedings.com